Universe, Neutrinos, Stars and Life

Dieter Frekers • Peter Biermann

Universe, Neutrinos, Stars and Life

Intriguing Insights from Astrophysical Research

 Springer

Dieter Frekers
Institut für Kernphysik
Universität Münster
Münster, Nordrhein-Westfalen, Germany

Peter Biermann
MPI für Radioastronomie
Bonn, Nordrhein-Westfalen, Germany

ISBN 978-3-662-70728-9   ISBN 978-3-662-70729-6   (eBook)
https://doi.org/10.1007/978-3-662-70729-6

Translation from the German language edition: "Weltall, Neutrinos, Sterne und Leben" by Dieter Frekers and Peter Biermann, © The Editor(s) (if applicable) and The Author(s), under exclusive license to Springer-Verlag GmbH, DE, part of Springer Nature 2023. Published by Springer Berlin, Heidelberg. All Rights Reserved.

This Springer imprint is published by the registered company Springer-Verlag GmbH, DE, part of Springer Nature.
The registered company address is: Heidelberger Platz 3, 14197 Berlin, Germany

If disposing of this product, please recycle the paper.

FOR THE READER

The authors of this book have tried to convey physical relationships as intuitively as possible and in an easy-to-read form. In doing so, they have largely dispensed with mathematical equations. Should mathematical formulas nevertheless occur in some places, these are specially marked in color (i.e., highlighted in yellow) and can be skipped without great disadvantage to understanding. The authors have also attempted to write the individual chapters in such a way that they can be read selectively and do not refer to contents of previous ones.

Münster, February 2021

The authors:

Dr. Dieter Frekers
Professor Univ. Münster, Germany
Fellow of the American Physical Society
Experimental nuclear and particle physics

Dr. Dr. h.c. Peter Biermann
Professor(apl) Univ. Bonn, Germany
Max-Planck-Institut für Radioastronomie
Adjunct Professor Univ. of Alabama, Tuscaloosa, USA
Honorary Doctor of Univ. Bukarest, Romania
Theoretical astrophysics and cosmology

CONTENTS

Chapter 1

Foreword

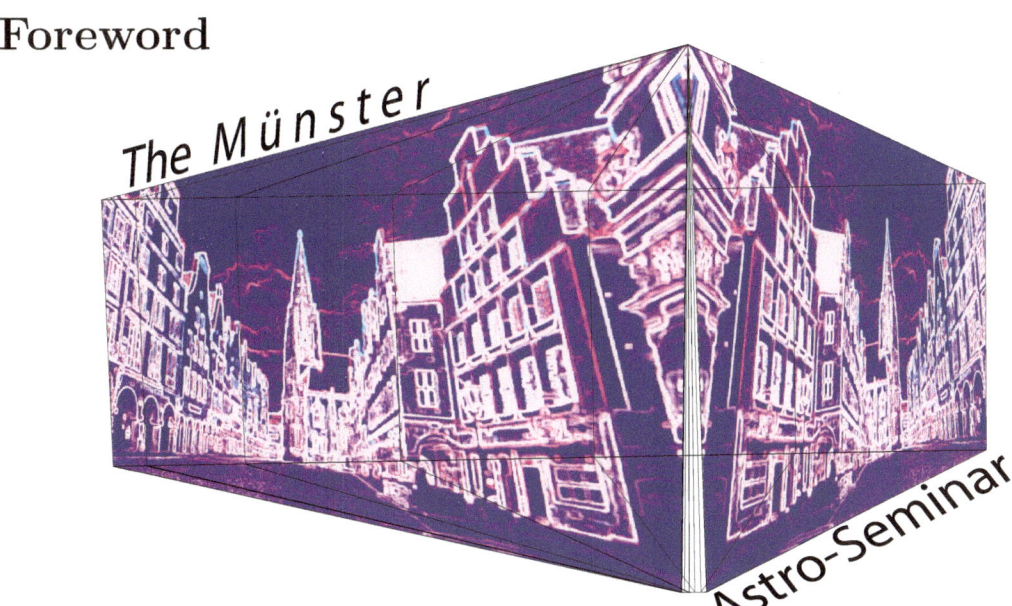

1.1 This is how it all began

The "Münster Astro-Seminar" is an event that has been held annually for more than 20 years, always at the end of September/beginning of October from Friday to Saturday and at the Physics Department of the University. The target group is an interested audience from all walks of life — and of course, the admission is free. During these two days, the current research and the latest results on specific topics from the fields of space research, astrophysics, astro-particle physics and cosmology are being presented in lectures and discussions by a few selected experts and scientists from these fields, and this always in a scientifically comprehensible and even in an enjoyable and entertaining way.

D. Frekers, P. Biermann, *Universe, Neutrinos, Stars and Life*, https://doi.org/10.1007/978-3-662-70729-6_1

The fact which makes this event even more special and unique is that it is organized down to the last detail by PhD students from the Institute for Experimental Nuclear and Particle Physics. They alone define the topic and arrange the program over the two days. They invite as speakers internationally renowned scientists from the different research areas, apply for necessary financial resources, organize the publicity campaigns, provide the premises and the necessary technical equipment, arrange for refreshments and small snacks during the breaks, and they usually also participate by giving short talks about their own work on a particular and matching topic. There is no interference of any kind on the part of the university lecturers and professors. The "Münster Astro-Seminar", which is by now firmly established in Münster, is in this way unique in all of Germany and a real success story, as over the years the event has attracted an ever increasing audience from an ever larger periphery of Münster.

In 2000, the Astro-Seminar started quite modestly with a lecture on an astrophysical topic by one of the authors of this book, Prof. Dr. Dr. h.c. Peter L. Biermann, member of the Max Planck Institute for Radio Astronomy and of the University of Bonn. The occasion was an event in Münster organized for scholarship holders of the German Academic Scholarship Foundation ("Studienstiftung") [1]. Afterwards, the host from the Institute of Planetology at Münster University, Prof. Dr. Elmar K. Jessberger, invited the attendees to his home, where a small remaining group of a handful of students from different departments discussed with the two colleagues further subjects of astrophysics and cosmology until late that night. The students were deeply impressed, and they proposed to repeat the event the following year in the same form and with an even broader range of topics.

Thanks to a young medical student from this small circle, this suggestion was indeed not forgotten, and the following year the group took the initiative and put together a program of topics and speakers to which the authors of

---

[1] The mandate of the non-profit foundation "Studienstiftung" (German Academic Scholarship Foundation): The German Academic Scholarship Foundation promotes students whose talent and personality give rise to expectations of special achievements in the service of the general public (translated text passage: https://www.studienstiftung.de/studienfoerderung/). In this context, events such as the one described here are promoted in particular.

this book were invited for a series of lectures on a freely chosen astrophysical topic. The number of interested students increased, although in the first few years the participants continued to be limited to the members of the above mentioned Scholarship Foundation. A subsequent survey among the participants showed that there was great interest in continuing this form of event in the years to come and also in making the event more public by expanding the circle of participants. Further annual events followed, which were again initially organized by the fellows of the Scholarship Foundation. Specially designed advertising posters were placed in the university buildings in order to draw the attention of students from all disciplines to this event. This quickly caused an increase of the number of participants to about 20 – 30. Appropriate rooms had to be booked, which was not easy because this "Astro-event" was still more or less private. However, the organizers were quite imaginative; there was always someone who knew this or that professor or caretaker who could be asked to make the premises (and the keys) available from Friday afternoon to Saturday evening. In the beginning, they met in a small lecture hall of the Anatomy Department, later e.g., in the seminar room of the "Pfarrzentrum Heilig-Kreuz" ("Parish Center of the Holy Cross") in Münster.

The awareness of the Astro-Seminar started to grow and the event got a new format with new faces in the course of time. In 2005, the event was held for the first time at the Institute of Nuclear Physics. Meanwhile, the organizers were students from this institute who, with great ambition and personal commitment, succeeded in reaching an audience even outside the university and attracting scientists from the international community to this event. The number of visitors unexpectedly skyrocketed and the seminar room in the Institute of Nuclear Physics holding about 35 seats soon proved to be much too small. In the following years the large lecture hall of the Physics Department together with the support from the technical staff could be obtained, and this also thanks to the special and unhesitant support of the Dean of the Department, who was able to even provide some extra financial support.

Surprisingly, the attractiveness of the event has not faded over the years, which was of course due to the evermore exciting new topics, but also because each time special highlights were built into the program, e.g.,

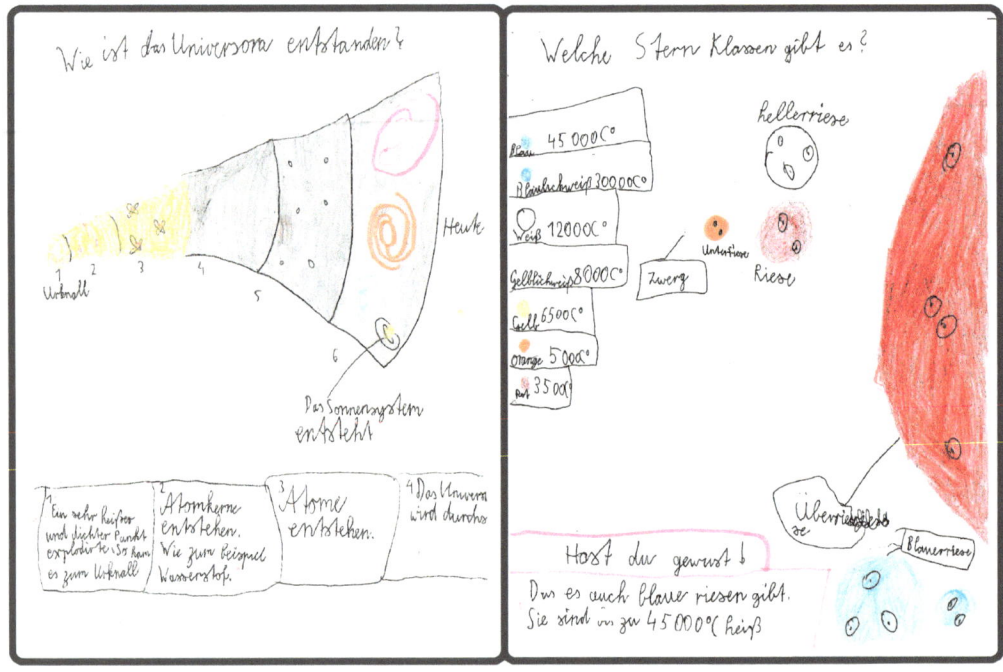

Fig. 1.1: The World in the eyes of an 8-year old.

a lecture on "Life in space" by ESA[2] astronaut Dr. Gerhard Thiele, a theatrical interlude about Victor Franz Hess, the discoverer of cosmic rays, or a particularly amusing lecture on "fact-vs-fiction" physics from the television series Star Trek. Small prizes for particularly good and sharp questions were regularly awarded after each lecture, laboratory tours were part of the program, or one was simply a point of contact for students who were considering studying Natural Sciences.

The student Andreas M. at the age of 8 was probably the youngest participant of the Astro-Seminar. Some lectures had impressed him so much that he independently designed a small series of pictures and passed them on to the authors of this book. Two of his drawings are included in this book (Fig. 1.1) to show that physics is an extraordinarily inspiring and exciting field, and this for all age groups.

Currently, the Astro-Seminar can count on $300 - 500$ visitors every year

---

[2] ESA = European Space Agency

(see Fig. 1.2). Congratulations to all the students who contributed to this tremendous achievement!

This book is an attempt to review the topics that have been touched upon in the years since the Astro-Seminar came into existence. However, it is not possible in hindsight to summarize or comment on the many lectures individually. The selection of topics in this book is therefore based on the subjective and very limited point of view of the authors.

The book begins (after clarifying what physics "is", and what it "is not") with an outline about Cosmic Inflation and pursues the question of how compelling such an amazing phase of development is that began immediately after the "Big Bang" and was already completed within an inconceivably short time of $\approx 10^{-30}$ seconds. After that it dives into the phase of the Early Universe and Early Element Synthesis, which started about some microseconds after the "Big Bang" and lasted approximately 10 minutes. The book will then illuminate how cosmic microwave radiation came into existence, and what it may tell us today; it further explains how elements are formed in supernova explosions, and how such explosions, which can outshine an entire galaxy for a few days, can be used to calibrate the expansion velocity of the Universe, thereby providing us with an extremely surprising (and ultimately Nobel Prize-winning) result. Chapters and sections about relativity, neutron stars, Black Holes, dark matter and neutrinos will follow, whereby also the questions will be put forward: «*how could Life arise*», and «*do neutrinos, which are at the 'edge of existence'*» have anything to do with it? Finally, we turn to cosmic rays. Cosmic rays are messengers of the past, but what do they tell us and how can one decode their messages?

Since man hasn't been witness to any of these long gone epochs of the Universe or to any of the processes that occurred at these infinite distances [3], the legitimate question arises: «*how does one know if no one has ever been there?*» Fortunately, these past and distant events have left a multitude of unmistakable fingerprints containing information which just needs to be decoded. From all of this, a consistent overall picture slowly comes to light, much like putting together a big puzzle with a still growing number of

---

[3] The term "infinite" is not to be understood here in the mathematical sense.

pieces. Interestingly, these studies also allow making short- and long-term predictions, e.g., for the solar system, for the galaxies, or even for events which lie in the far distant future of the Universe.

However, in order to understand that "Physics" is not a collection of intellectual fantasies born in the minds of scientists, the authors of this book think it is important to first provide some clarity about "what physics is and what physics is not".

Fig. 1.2: Photo-reminiscences of the Astro-Seminar in the large lecture hall of physics

Chapter 2

Physics and Cognizance

2.1   Physics is — what ... ?

Yes — what is it...? There is a seemingly endless number of more or less profound and deeply rooted philosophical, theological or other meaningful discourses, which attempt to subsume the essence of natural sciences and the insights arising from it in some suitably matching scheme of thought. It will not be the objective of this book to take stock of this. Instead, the authors will try to draw a simple and common sense picture, freed from

9

D. Frekers, P. Biermann, *Universe, Neutrinos, Stars and Life*, https://doi.org/10.1007/978-3-662-70729-6_2

theological and philosophical burdens. This is in keeping with the daily work of natural scientists, who merely deal with the simple question of how things are put together and how they interact with each other. By following this line one quickly comes to realize that Nature is knitted up in an utterly simple, transparent and rational manner, whilst at the same time maintaining clear and strict laws. And with a little intelligence, these can be combined and utilized in an immensely beneficial way[1].

During the Middle Ages and even up to Modern Times, trying to understand and to explore Nature became exactly a bone of contention, at least in the Christian-European cultural sphere: the study of the laws of Nature and exploring the essence of Nature was classified as heretical and by no means in conformity with divine teachings[2]. However, Nature reveals itself voluntarily to us human beings, all by itself and without any of our doing. Under no circumstances does a human peer into the cards of an "Intelligent Being", let alone in an improper or forbidden way.

Furthermore, we will even see that processes that took place far in the past (e.g., shortly after the "Big Bang", 13.8 billion years ago) had and still have an enduring effect on the evolution and the composition of our present world, and are also of continuing relevance for the world's future development. This makes it possible to trace back processes from bygone epochs in all detail by means of the laws laid down by Nature. The basis for this is the creation of robust and resilient description models laid out in a consistent mathematical language. These models must be inferred solely by observing recurring patterns and must be fundamentally falsifiable from the design stage, i.e., be verifiable through experimental tests (which is equivalent to "observing"). This includes, foremost, that models are capable of making verifiable predictions using mathematical inference and deduction. Models which do not provide predictions or which describe or hypothesize things that do not exist, are not falsifiable and therefore useless. An example: A model that describes the era before the "Big Bang" is of no relevance if it does not at the same time specify in mathematical language

---

[1] A trivial example: Force creates a change in motion (the law), which allows one to move unerringly from A to B (the benefit).

[2] The reader may want to consult the biography of the last Staufer emperor and king Frederic II. (1194–1250) and his astoundingly modern view of how to uncover the laws of Nature by making observations and carrying out experiments.

the measurable properties of this era. The scenarios captured in this book about the evolution of our Universe, though some sound truly spectacular, are therefore very pragmatically and unemotionally based on known laws of physics and well established models and are not figments of scientists' imaginations.

Fig. 2.1 summarizes in a simple way how to establish physical models that describe the laws of Nature. In words:

1 start with observations,

2 find regulatory patterns and create a model for their description in a mathematical language,

3 make predictions on the basis of the developed models,

4 test the predictions and check the validity of the model,

5 improve (e.g., by further observation and with better techniques) the model and start again with point 2.

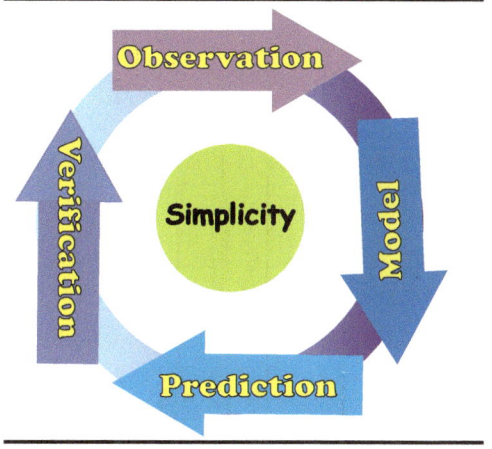

Fig. 2.1: The essence of physics

A central feature is the simplicity of a model, i.e., its restriction to a minimum number of relevant parameters. It may be instructive to remind oneself of the large number of the naturally occurring chemical elements together with their stable isotopes (approx. 250), which are ultimately composed of only 3 different building blocks, the proton, the neutron and the electron.

Furthermore, the power of a descriptive model is given by its ability to make correct and verifiable predictions. In this way, for example, Quantum Mechanics or the Standard Model of elementary particles have established themselves by having been able to withstand all experimental tests with an astounding accuracy (at times up to the 15th decimal place).

2.2   Answer: Physics is — this!

In oder to better understand what the very essence of physics is, we will in the following section turn the argument around and examine what physics is NOT and what physics does NOT achieve under any circumstances. These explanations are important for the understanding of this book, in order to keep a critical view of the processes that happened in the past or synonymously at great distances. The individual and essential arguments are:

1 Physics merely describes and does NOT explain
2 Physics does NOT claim ultimate Truth
3 Physics does NOT provide Proof of Truth
4 Physics is NEVER exact
5 Physics does NOT describe things that never
   appear (i.e., things that do NOT exist)

In the perception of natural sciences, and of physics in particular, the general public frequently seems to be unaware of these important features, an unawareness that at times leads to completely absurd chains of arguments. Yet, some self-criticism may be in order, because the reason this happens is also due to the at times imprecise and lax language of physicists and scientists, when they talk about «*proof*» but mean «*verification*» or talk about «*exact and true*» but mean «*accurate within the scope of the measurement accuracy*». In particular, the term «*within the scope of the measurement accuracy*» ought to be noted! We will come back to this later.

This may all sound a bit subtle and philosophical, so at this point a little more explanation of the individual statements is in order.

▶ **1: Physics merely describes and does NOT explain**

The laws of Nature, which manifest themselves foremost in the area of physics, are the basis for any being, especially for any intelligent being, to create a plan, which allows it to find its way around in Nature. Even a toddler will make such a plan in order to understand and coordinate motion. He/she does this effortlessly and without mathematics, just by experimenting, observing and communicating. Mathematics comes into play later, and with the help of it physical processes can be described through models. Mathematics is therefore only a tool, or, if one wishes, the language in which models/laws are being communicated or logically linked with one another (e.g., distance covered = speed multiplied with time passed). Physics does not claim to explain anything, it merely describes those things in life, which are relevant for being able to make verifiable and reliable predictions. An explanation of why things are the way they are and not otherwise is not important to the plan. What is important, though, is that a model fits into a self-contained, uniform and simple overall model of Nature. One may put it in different way, for physics Nature appears simple and regular, and it is utterly nonphysical to construct separate models for each rainbow in the sky or for each shooting star appearing at night. Those types of particulate models are useless for predictions and planning. «*Miracles may happen*» they say, but they are not within the domain of predictable physics laws, though they are not negated by physics either. There is no mandate for that. Physics is not in conflict with Religion and Belief.

▶ 2: Physics does NOT claim ultimate Truth

It is also not of importance that a description model is ultimately correct and true. The only thing that matters is that it can make reasonable predictions within the context of an accuracy which is relevant for enabling progress and moving ahead, and/or making a meaningful planning possible. This stays true up to the point when weaknesses of the model become apparent, which then require improvements or extensions. As a simple example, one may take the repeatedly used legend of the apple, which allegedly fell on Isaac Newton's head in 1666. From this apple anecdote (no! — it was the "Philosophiae Naturalis Principia Mathematica") the theoretical model of Newtonian (or classical) mechanics unfolded, which became the basis of all motion processes on Earth (and beyond). It describes these processes with a remarkable precision, verified by an almost infinite number of experimental tests. And if, for instance, Bernoulli's fluid mechanics is added to Newton's law, then even airplanes can be made to fly.

In this way, Newtonian mechanics seems to represent the ultimate truth — unfortunately badly wrong! With the development of techniques allowing for a continuous increase of experimental precision over the last $50-100$ years, more and more discrepancies between measured quantities and classical predictions emerged. A remedy for these was finally provided by the mathematically quite demanding theory of General Relativity, which is most closely associated with the name Albert Einstein. It was developed solely to account for the experimentally observed constancy of the speed of light. In fact, General Relativity elegantly reduces to Newtonian mechanics in the limiting case of a so-called «*small laboratory*», which also means, it is not a particulate model. A «*small laboratory*» is realized whenever the curvature of space-time in this «*small laboratory*» environment has a negligible effect on the relevant measurement results or on the conception of strategies. This is indeed the case for practically all motion processes on Earth. If highest precision is required, e.g., for the Global Positioning System (GPS), for the determination of a time standard (important for digital radio communication, i.e., for mobile phones), or for satellite navigation, then the curvature of space-time caused by the presence of masses and their different distributions is significant and must be taken into account.

The constancy of the speed of light is not intuitive in the simple human way of thinking, but it is undeniably observed as such. The accuracy with which the theory of General Relativity stands up to experimental scrutiny is impressive. To date, there is no indication at which precision or after which decimal point this theory ought to be replaced by an even more extensive or even more general theory. The question arises: is this theory now after all the true and correct theory, or the other way round, at which decimal point does Truth begin? The answer is superfluous because the theory of General Relativity has at least one unresolved and inherent and most worrisome shortcoming. It is divergent near a Black Hole, i.e., infinities occur at the boundary to the Black Hole. The same happens at the time of the "Big Bang". Infinities, however, are simply and utterly nonphysical. At the (fictitious) transition from a finiteness into an infiniteness and vice versa, causality is violated, i.e., time sequences are thrown into disarray, or more pointedly, they are completely dissolved. Everything is and remains infinite forever with no beginning and with no end.

Of course, there are other disciplines in physics, such as electrodynamics (keywords: electricity, television), thermodynamics (keywords: temperature, life), hydrodynamics (keywords: weather, airplane), and especially the theory of quantum mechanics (keywords: atom, computer). They all intertwine in an elegant and consistent way and describe natural processes in close interaction with each other, and they do this with astonishing precision. Based on this interplay, even more distant fields such as chemistry, biology or medicine were able to manifest themselves as independent disciplines over the course of time. They all have in common that they merely describe natural processes, at times not even deterministically but only on the basis of probabilities. In physics this probabilistic situation is most notably evident in the field of thermodynamics and even more so in quantum mechanics. Truth, however, cannot be subject to probability. The fact that quantum mechanics does not describe physical phenomena deterministically must simply be accepted. Of course one may ask, does Nature on the smallest scales gamble with its own laws, or is it perhaps because on these scales the usual definitions of time and space no longer apply? So far, there is no answer to this question. Regardless of this, quantum mechanics passes all tests; it thus establishes itself as an extremely useful descriptive model.

▶ 3: Physics NEVER provides Proof of Truth

This is perhaps the most important point about what physics is NOT, seeing that this is frequently ignored and disregarded in numerous pertinent discussions and likewise in numerous philosophical treatises.

Physics can NEVER lead to proofs of truth, this is simply impossible. Physics, or synonymously the knowledge and the progress that develops from it, are always and without exception connected with measurements. However, every outcome of a measurement is subject to an error margin. A measurement without an error is inconceivable. Conversely, providing a result from a measurement without an estimate of the measurement error is also useless. Every technical drawing requires tolerances be specified. Without this information, no technician or engineer will ever begin to manufacture a workpiece. In daily life, though, not every trivial piece of information like e.g., the temperature of the day, the size of an area, the speed of a car, the energy used in private households, the indications on food containers or whatever else one may imagine, ought to be provided with an error statement — that would be highly nonsensical in daily life, yet in the technical and the scientific fields this information is essential.

One can formulate the above statement 3 even more pointedly by means of an example. In the ideal world of mathematics we learn (mostly in school) that parallels intersect at infinity. One may ask if there is any teacher who has experimentally verified parallelism even over a length of only 10 meters? Obviously, this seems to make so much intuitive sense that one would not even have to bother checking it. Carl Friedrich Gauss $(1777-1855)$, to whom we will come back later, had already doubted this dogma for the real world, but unfortunately his measuring accuracy was not sufficient to detect deviations.

Furthermore, in the ideal world of mathematics we learn that the circumference of a unit circle is 2π. In the real world, because of the infinite series of the number π, this result is not at all provable by any measurement and certainly not correct either. The mere presence of a mass inevitably results in a curvature of space, which on Earth already leads to a discrepancy in the 10th digit after the decimal point — a giant effect, if one considers how many digits still follow, but completely negligible for most things in our «*small laboratory*» Earth. And now, one may imagine how many physics formulae contain such transcendental numbers such as the number π or Euler's number e, or for instance those irrational numbers expressed by simple square roots.

▶ 4: Physics is NEVER exact

This is almost synonymous with the above statement 3. Physics always takes place in a real world. And mathematics is the language, in which physics (within the framework of theories and models) is communicated and quantified in a uniform and generally understandable way, although mathematics always and with no exception describes physics in an idealized environment. Yet, progress in the real world of physics is inconceivable without the language of mathematics, and in fact, physics is even a driving force behind the steady advancement of mathematics. Mathematics sees itself as an exact science, and if one leaves epistemic subtleties aside, this is a reasonable assertion. The logic of mathematics is always based on certain presuppositions, and these are always and without exception prefixed in mathematical conclusions (i.e., proofs). The presupposition made in the present case that the circumference of a unit circle is 2π is Euclidean geometry or the «*flatness*» of space. Of course, the circumference of a circle can be mathematically calculated in any curved geometry, but it does not change the fundamental inability to quantitatively prepare or single out a very specific mathematical presupposition and then claim this to be the ultimate truth by means of an inherently error-prone measurement.

▶ 5: Physics does NOT describe things that do NOT exist

This is in itself an amusing assessment and, of course, self-evident. How should one describe things that don't exist? What knowledge could be gained, and above all, which models could be created from this? Which verifiable predictions are these supposed to make? Which technological progress could be derived from describing things that don't exist in any way whatever? Nevertheless, there are always scurrilous arguments that attempt to discredit scientific research as allegedly incapable and lacking knowledge. Of course, physics can never prove that there is NOT an invisible [3] teapot orbiting the Earth. To cite this as an argument for its existence, is scurrilous and absurd. Of course, science can never prove that the protective effect against measles or chickenpox does NOT exist if lying down on a meadow at midnight under a full moon. Conversely, to predict such a protective effect simply because one hasn't yet contracted chickenpox, or one is convinced of it, is also absurd and unrealistic. There are many variants of this line of arguments. They represent a turning away from the firm laws of Nature and take us into abstruseness and back into long gone times of mysticism and superstition [4].

There is no question that so-called "zero"-experiments are a part of physics, however, these ought not be contrived as an attempt to prove something does NOT exist. Perhaps the best known experiment of this type is that of Michelson and Morley (1881 and 1887), which is being repeated even to this day with increasingly refined techniques. The aim was, and still is, to determine the motion of Earth relative to a hypothetical ether that acts as the carrier of light-waves. The light-speeds in and against the direction of Earth's motion should be different in the presence of such a carrier. All

---

[3] In the sense of "not at all interacting".

[4] Here we are not talking about placebo effects; they are scientifically quite well-founded.

experiments measured a "0"-value, but be aware about: "within the error margin of the measurement !" Today, these experiments are considered a test of the so-called Lorentz invariance, i.e., a test of the constancy of the speed of light in a uniformly moving system. At present, relative deviations in the speed of light $\Delta c/c$ in the order of 10^{-17} can be ruled out experimentally, still in agreement with the theory of relativity, which predicts the value "0" for this [5],[6].

---

[5] Ch. Eisele, et al., *Laboratory Test of the Isotropy of Light Propagation at the* 10^{-17} *level*, Physical Review Letters 103, 090401 (2009).

[6] S. Herrmann, et al., *Rotating optical cavity experiment testing Lorentz invariance at the* 10^{-17} *level*, Physical Review D 80, 105011 (2009).

Chapter 3

Measured Values $\longleftrightarrow$ Magnitudes

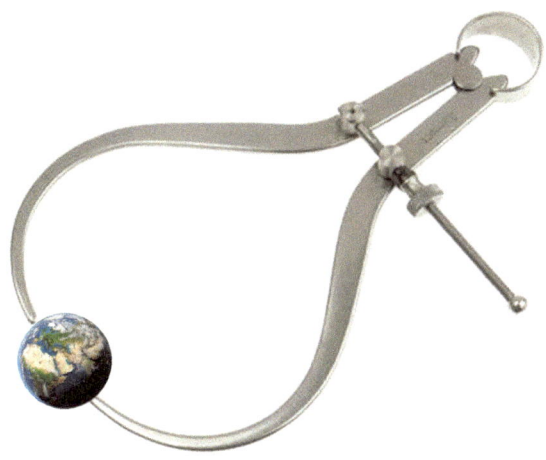

In this book we will frequently talk about measurements and the various dimensional units given to the outcome of these, i.e., their results. Dimensional units are fundamental if physics processes are to be quantified and compared with each other. Moreover, in astrophysics and cosmology the scales or the range of magnitudes play a decisive role, and frequently comprehension seems to fade when trying to map cosmological quantities onto our small personal domain of life. This remains a perpetual challenge even for scientists, who deal with these issues professionally on a daily basis.

© The Author(s), under exclusive license to Springer-Verlag GmbH, DE, part of Springer Nature 2025
D. Frekers, P. Biermann, *Universe, Neutrinos, Stars and Life*, https://doi.org/10.1007/978-3-662-70729-6_3

3.1 Distances

These are usually expressed in units of meters (m) or kilometers (km). For cosmological distances, these units are generally regarded as not particularly useful, and one regularly resorts to light-years (ly) or parallax-seconds (pc), which are the standard dimensional units in that case.

Meter and kilometer are perceivable units in everyday life. Dealing with a light-year, which is the distance light travels within a year, the connection to personal life dwindles, especially if one were to express this distance in kilometers. Here a simple conversion of a light-year to kilometers:

$$1 \text{ year} \simeq \pi \times 10^7 \text{ s}$$
$$\text{speed of light} \simeq 300,000 \text{ km/s}$$
$$\text{therefore: } 1 \text{ light-year} \simeq \pi \times 10^7 \times 300,000 \simeq 9.42 \times 10^{12} \text{ km} \simeq 10^{13} \text{ km}$$

Surely, a year does not exactly translate to $\pi \times 10^7$ seconds, because a year really doesn't have anything to do with π. Nonetheless it's a relatively easy number to remember. The correct number is $3.155\,760 \times 10^7$ seconds, but note that this value makes exactly 365.25 days to a year. This definition of a year was set by the International Astronomical Unit (IAU) to avoid making small annual adjustments resulting from changes in the Earth's orbital period. Similarly, the speed of light is not exactly 300,000 km/s either, but rather 299,792.458 km/s [1]. In this way, the exact value of a light-year is $9.460\,730\,472\,580\,800 \times 10^{12}$ km, which is an awkward number to remember if one ever wishes to make quick estimates. The difference between this value and the easier to remember number of $(\pi \times 10^7) \cdot (3 \times 10^5) = 9.42 \times 10^{12}$ km, is just 1.5 light-days, which is about the size of our solar system. Present day precision measurements of distances smaller than about 1000 ly do much better than this, whereas at larger distances (above about 10,000 ly) this small difference as well as the 5% discrepancy between a light-year and the even more generously estimated value of 10^{13} km gets to be less and less

---

[1] For practical reasons the year is fixed by the International Astronomical Unit (IAU) to 365.25 days. Further, the quoted value for the speed of light comes by definition and carries no experimental error — the reason one chose this unwieldy number rather than the much easier to remember 300,000 km/s, is a different story, as this would have required a significant adjustment to the length scale of a meter.

significant.

The parallax-second (pc) originates from astronomical observations and is obtained by simple trigonometry. It is defined as the distance at which the Earth's orbital radius around the Sun, which is 149.598 million kilometers or one astronomical unit, subtends an angle of one arcsecond (i.e., $1''$ is 1/3600 of an angular degree). Conversely, a fixed star at this distance, when observed from Earth, seems to move by an arc-size of $1''$ over the course of half a year. The conversion to light-years gives:

$$1\,\mathrm{pc} \simeq 3.2616\,\mathrm{ly} \simeq 30.9 \times 10^{12}\,\mathrm{km}$$
$$\text{with: } 1000\,\mathrm{pc} = 1\,\mathrm{kpc},\ 1000\,\mathrm{kpc} = 1\,\mathrm{Mpc},\ 1000\,\mathrm{Mpc} = 1\,\mathrm{Gpc}$$

This angle can also be mapped to a more easily imaginable reference size, which is the diameter of a human hair at a distance of approximately 20 meter. Today's astronomical instruments (e.g., the Gaia space probe) have an angular resolution corresponding to the diameter of a human hair at a distance of approx. 2 km or $0.01''$. Distances of stars up to about 100 pc ($\simeq 300\,\mathrm{ly}$) can be determined by simple triangulation, and with a few tricks and for selected objects even up to 5000 pc. The closest star Proxima Centauri has a parallax of only $0.772''$ (corresponding to a

Fig. 3.1: An artist perception of the Milky Way Galaxy

distance of 4.225 ly or 1.295 pc), which at this point already shows that the area of 300 ly around our solar system is only a tiny fraction of the entire Milky Way Galaxy. In the visible spectrum the Milky Way has a diameter of about 200,000 ly ($\simeq 60,000\,\mathrm{pc}$), in the outer part a height of about 3000 ly ($\simeq 900\,\mathrm{pc}$) and in the inner part a height of about 20,000 ly ($\simeq 6000\,\mathrm{pc}$). Some 200 billion stars cavort in this volume. In its center ($\simeq 27,000\,\mathrm{ly}$ away from the solar system) resides a Black Hole (Sagittarius A*) with a mass of about 4.2 million solar masses and a diameter of 25 million km, which in size is relatively small, as it translates to only about 83 light-seconds.

In order to be able to determine distances of more distant objects as well as those located outside the Milky Way with some certainty, one has to resort to other methods. We will meet these later in this book.

3.2   Energies

These are measured in watt-seconds (Ws), joules (J) and electron-volts (eV).

Here, watt-second and joule are one and the same. Frequently one also encounters the newton-meter (Nm) as a unit for energy or work, but the newton-meter is more readily used as a unit for torque, though one ought to be careful: torque as a vector quantity is not identical to the scalar quantity energy, despite identical units.

In the field of cosmology, one rarely encounters the unit watt-second. However, if one deals with large and compact objects and their movements (e.g., planets, stars/suns, galaxies), then watt-seconds and joules (or sometimes even the outdated unit erg $= 10^{-7}$ Ws) are more common, perhaps because they can be mapped more easily to things that we encounter in everyday life (e.g., electricity consumption is billed in kilowatt-hours (kWh)). To make a comparison of the orders of magnitude, we take a meteorite of around 500,000 tons (i.e., such an object still has a rather modest diameter of $50 - 100$ meters). Objects of this size collide with the Earth every 1000 years on average. If these objects hit the Earth from the side facing away from the Sun, then they typically have velocities of approx. $50 - 60$ km/s, so that energies of about 170×10^9 kWh are released during such an event. Taking countries like Germany or the US, this corresponds to about 5%, resp. 0.7%, of their total energy consumption in 2021 (for 2021 given as 3.4×10^{12} kWh, resp. 28.5×10^{12} kWh [2],[3]). However, there is a decisive difference: in one such collision, the energy is released in a few seconds. One can now work

---

[2] Bundesministerium für Wirtschaft und Energie, Federal Ministry for Economic Affairs and Energy

[3] U.S. Administration and Information Administration

Tab. 3.1: Scenarios of impacts by compact objects on planet Earth, adopted and adapted from: *The Impact Hazard*, Essay von David Morrison, Clark R. Chapman and Paul Slovic in *Hazards due to Comets and Asteroids*, University of Arizona Press, 1994. The indicated densities relate to stone meteorites ($\approx 3\,\mathrm{g/cm^3}$) and iron meteorites ($\approx 5\,\mathrm{g/cm^3}$), the latter having a higher probability of reaching the Earth's surface in an almost compact form.

Impactor properties: density $3-5\,\mathrm{g/cm^3}$, collision speed $\approx 50\,\mathrm{km/s}$ 1 Megaton (Mt) TNT equals an energy release of 1.16×10^9 kWh				
diameter in meter	mass in Mt	energy in Mt TNT	frequency in years	consequences
< 50	< 0.2	< 10	< 1	meteors burn up and shatter, usually do not reach the surface of the Earth
75	$0.6-1$	$10-100$	1000	land impacts destroy areas of typical large cities
160	$6-10$	$100-1000$	5000	land impacts destroy areas of typical metropolitan cities (Berlin, Tokyo)
350	$60-100$	10^3-10^4	15,000	land impacts destroy areas of small states, ocean impacts produce smaller tsunamis
750	$500-1000$	10^4-10^5	66,000	land impacts destroy areas of medium-sized countries (Ireland), ocean impacts produce large tsunamis
1,700	$5000-10,000$	10^5-10^6	250,000	land impacts destroy areas of large states and countries (California, France), result in long-term atmospheric effects.

out what orders of magnitude of energy one will be confronted with, when two neutron stars, each having the mass of the Sun, collide with a speed that may be a good fraction of to the speed of light.

Table 3.1 summarizes possible scenarios of meteorite impacts of various sizes along with their estimated rate of occurrence. To give an idea of these numbers, some of the most famous meteorite impacts are briefly outlined.

The Chelyabinsk meteorite struck on February 15, 2013 at around 9:20 a.m. local time and was an event that was captured by a large number of video recordings. From the destructive energy it was possible to deduce its mass at around 12,000 tons and its diameter at around 20 meters. It turned out to be a stone meteorite with an average density of approx. $3.3\,\mathrm{g/cm^3}$. It hit the Earth from the side facing the Sun at a relatively low speed of approx. $20\,\mathrm{km/s}$. The end result: 3700 damaged buildings due to the pressure wave and approx. 1500 injuries.

The Tunguska event on June 30, 1908 in what is now the Krasnoyarsk territory was also likely due to a stone meteorite impact. The approximate diameter of the meteorite could have been in the order of $30-80$ meters. Since there was no crater, this object must have blown up in the higher atmosphere (i.e., around 10 kilometers). This scenario was derived from the (albeit few) eyewitness reports as well as from the devastation wrought over an area of almost $2000\,\mathrm{km^2}$.

A conclusive overall picture of the sequence of this event can still not be put together to this day.

The Barringer Crater in Arizona (Fig. 3.2) goes back to a meteorite impact, which occurred about 50,000 years ago. The meteorite had a diameter of approx. 45 meters, weighed about 300,000 tons and consisted primarily of iron. It hit the Earth at a speed of about $15-30\,\mathrm{km/s}$ and left a crater $1.2\,\mathrm{km}$ wide and $170\,\mathrm{m}$ deep.

Fig. 3.2:　The Barringer crater

The meteorite which created the Nördlinger Ries (Fig. 3.3) in what is today the area between the Swabian and Franconian Jura (Germany), hit about 14.6 million years ago and likely had a diameter of around $1.5\,\mathrm{km}$ and a speed of around $50\,\mathrm{km/s}$. The crater it created on impact was about $25\,\mathrm{km}$ in diameter and about $500\,\mathrm{m}$ deep.

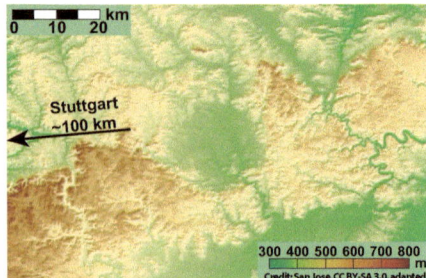

Fig. 3.3:　The Nördlinger-Ries crater

The Chicxulub crater (Fig. 3.4) near the Yucatan peninsula in Mexico is undoubtedly evidence of one of the most spectacular meteorite impacts in Earth's latest history, resulting in the most far-reaching consequences for the evolution of all living species at that time. The impact date has by now been determined with high precision to 66.04 million years ago. Given today's knowledge, the event led to the abrupt end of the 200 million years

Fig. 3.4: The Chicxulub crater

lasting era of the dinosaurs. But even more, it caused a sudden and severe global climate change, which led to a general extinction that probably affected 70–75% of all living species on Earth. These climatic aftereffects lasted for about 100,000 years. Initially, there were decades of global winter with a temperature drop of almost 30° C. The global winter combined with a long period of darkness caused by the enormous amount of micro-particles that were injected into the entire atmosphere or even thrown into orbit, resulted in an almost complete cessation of photosynthesis. After all the debris slowly started to fall back to Earth's surface and the days gradually cleared up, the vast amounts of CO_2, which had been generated by the impact and the ensuing fires, led to a subsequent greenhouse effect [4),5),6)]. The crater depth caused by the impact must have been about 30 to 35 km. Together with the crater diameter of 166 km, the dimensions of the meteorite or asteroid could eventually be estimated at 10–15 km in diameter and the impact speed at 20–40 km/s as well as the impact energy at around 10^{24} J or a few 10^{17} kWh. This is roughly 1000 times the current world's annual energy consumption.

[4)] D. S. Robertson et al., *K-Pg extinction: Reevaluation of the heat-fire hypothesis*, Journal of Geophysical Research, Biogeosciences 118, 329 (2013).

[5)] J. Brugger et al., *Baby, it's cold outside: Climate model simulations of the effects of the asteroid impact at the end of the Cretaceous*, Geophysical Research Letters 44, 419 (2017).

[6)] One can ponder what the biodiversity of our Earth would look like today, had it not been for this impact. In view of the 200 million years lasting era of the dinosaurs, their rule would hardly have changed over a comparatively short period of another 66 million years until today.

These examples are intended to show that at astronomical and cosmological scales (which are not yet reached at all in the events described here) one quickly loses any perception of the powers of ten. Similar to the previously discussed distances, one could now assume that new energy scales will be defined that are more appropriate to the enormous orders of magnitude that occur. This would perhaps make it easier to relate them to simple pictures and concepts. Yet, the "electron-volt" as an energy unit goes exactly the opposite way and designates a scale, which seems to be more fitting to atoms and elementary particles. That this is still useful for understanding cosmological concepts will be highlighted in the next section.

Electron-volt:
The electron-volt (eV) is the energy a charged particle of one unit of charge (e.g., an electron) acquires when it traverses an electrical voltage drop of one volt (1 V). Since this energy constitutes kinetic energy (i.e., $E = \frac{1}{2}mv^2$), the definition can be applied as well to non-charged particles (e.g., neutrinos), which receive their energy via other, non-electrical processes.

The charge of an electron is $e_0 = -1.602\,176\,62... \times 10^{-19}$ ampere-seconds (As) (or coulomb, Cb), therefore, after passing through a voltage difference of 1 V, the electron has acquired an energy of $E = 1.602... \times 10^{-19}$ Ws (or J). In order to find one's way around this new and miniscule scale, roughly the following applies: In terms of energy, atomic physics processes lie in the eV – keV range, nuclear physics processes in the keV – MeV range, and elementary particle processes are typically in the GeV range and higher.

An electron starting at rest, which traverses a voltage difference of 1 V unhindered, acquires a velocity of about 600 km/s. An approximately 1800 times heavier hydrogen atom still has a velocity of about 14 km/s at 1 eV and a 100,000 times heavier iron atom a velocity of 1.8 km/s. These are velocities of about the same order of magnitude as they were already mentioned in the previous section. Moreover, it shows that the energy of one of the above mentioned meteorites is actually quite small on a microscopic scale (i.e., the energy per atom), and only by the collective effect, i.e., through the sum of the energies of all particles, whose number is well in the oder $10^{30} - 10^{40}$, the macroscopic or the cataclysmic effects come to bear. What is important here is the collective effect, because in the case of an ordinary gas like hydrogen or air at room temperature the molecules of hydrogen (H_2)

have already an average velocity of 1.9 km/s and those of air an average velocity of 0.5 km/s[7]. These only create the permanent pressure of the gas, and as long as the pressure is balanced, there is no damaging effect at all.

---

[7] The mean velocity $\overline{v} = \sqrt{\langle v^2 \rangle}$ follows from the kinetic theory of gases and can easily be calculated from the relation $\frac{3}{2}k_B T = \frac{1}{2}M\langle v^2 \rangle = \frac{1}{2}Mc^2\frac{\langle v^2 \rangle}{c^2} = \frac{1}{2}Mc^2\langle \beta^2 \rangle$ with the Boltzmann constant $k_B = 8.617\,33 \times 10^{-5}$ eV/K, the temperature $T = 295$ K (room temperature) and the mass of the molecule or atom $Mc^2 = A \times 931.494\,095 \times 10^6$ eV and A the mass number (e.g., $H_2 \simeq 2$, $N_2 \simeq 28$, $O_2 \simeq 32$, air $\simeq 28.8$). The conversion factor $931.494\,095 \times 10^6$ follows from the relativistic energy-mass relationship $E = mc^2$. The quantity β is the velocity in units of the light velocity c. With the help of the gas theory, the surface impact rates can be calculated as well, which is for hydrogen gas at room temperature and normal pressure around 6×10^{27} impacts per square meter and second, and for air around 1×10^{26} impacts per square meter and second.

Chapter 4

A Short Trip to the World of Elementary Particles

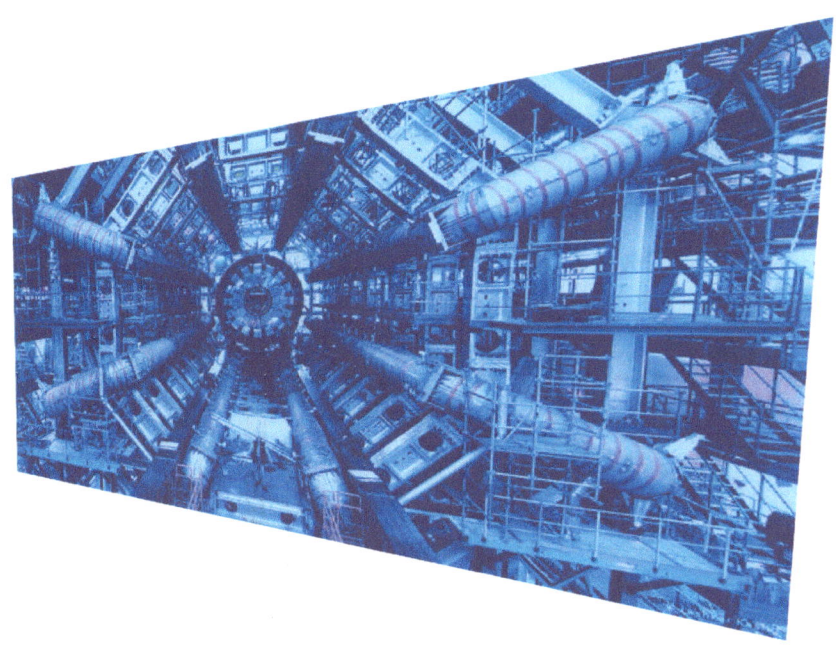

4.1 Quarks and leptons

From the mid-1970s onward, elementary particle physics experienced some of its greatest and most important discoveries for comprehending the forces of Nature and the internal cohesion of matter. At major experimental facilities such as DESY/HERA in Hamburg, SLAC at Stanford University

D. Frekers, P. Biermann, *Universe, Neutrinos, Stars and Life*, https://doi.org/10.1007/978-3-662-70729-6_4

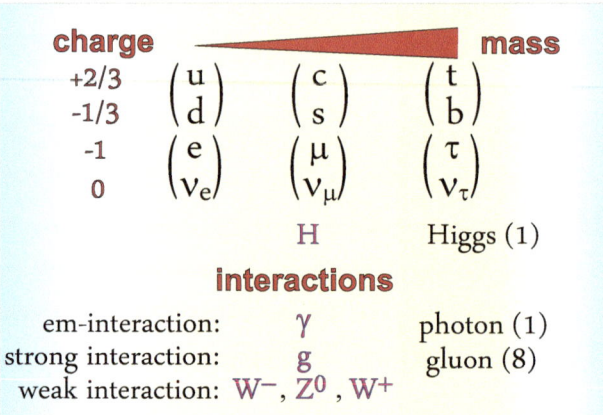

Fig. 4.1:   Elementary particle model with the 3 quark families and the corresponding 3 lepton families as well as with the interaction particles, which are the photon for the «*electromagnetic*», the 8 gluons for the «*strong*» and the 3 bosons W^+, Z^0, W^- for the «*weak*» interaction. In the diagram, H is the mass-generating Higgs particle. – The masses of the quarks and leptons increase from the left to the right. However, given the present experimental situation, there is still a possibility for ν_τ to be the lightest neutrino in this ordering.

in California, CESR at Cornell University in Ithaca (NY), or CERN[1] in Geneva, it was possible to break down the internal structure of nucleons and to visualize the "field particles" that mediate the «*strong*» and «*weak*» forces. This was a significant step closer towards understanding the interaction and structure of the Universe at the early time of its formation (see the following Chapter 5).

Figure 4.1 shows the basic building blocks of matter in a simplified form, as it is currently viewed and accepted. The so-called "hadronic" (primarily strongly interacting) matter is made up of quarks[2], while "leptonic" (primarily weakly interacting) matter is made up of the charged leptons (electron, muon, and tauon) and their uncharged partners, the neutrinos.

---

[1]   DESY **D**eutsches **E**lektronen-**SY**nchrotron, HERA **H**adron-**E**lektron-**R**ing-**A**nlage, SLAC **S**tanford **L**inear **A**ccelerator, CESR **C**ornell **E**lectron-positron **S**torage **R**ing, CERN **C**onseil **E**uropéen pour la **R**echerche **N**ucléaire.

[2]   The name "quark" was coined by Murray Gell-Mann (Nobel Prize 1969). Gell-Mann was an admirer of the Irish writer James Joyce. In the novel *Finnigans Wake*, the passage *three quarks for Muster Mark* gave him the inspiration for the name.

The quarks arrange themselves into three families (or generations), the "up"-type quarks consisting of the "up" quark **u**, the "charm" quark **c** and the "top" quark **t**, and the "down"-type quarks consisting of the "down" quark **d**, the "strange" quark **s** and the "bottom" or "beauty" quark **b**.

The "up"-type quarks carry the electric charge $+\frac{2}{3}$ and the "down"-type the electric charge $-\frac{1}{3}$, each in terms of the unit charge e_0. Thus, the quarks are also subject to the «*electromagnetic*» interaction.

The leptons e^-, μ^-, τ^- carry charge -1 (also in terms of the unit charge), while the corresponding neutrinos ν_e, ν_μ, ν_τ are electrically neutral. Each lepton family is associated with the respective quark family owing to the fact that decays and transitions are largely family-conserving.

To complicate matters a bit more: In addition to the electric charge $(+, -)$, quarks carry another, gluonic charge, which occurs in three variants and is classified, for lack of a better term, as the color-charge, i.e., red, green, or blue. Bound quark-states occur only color-neutral (i.e., "white") and, according to the classical color scheme, they must be built either from the three different "color" combinations or from a "color – anti-color" combination. The former are called baryons (3-quark systems, e.g., proton, neutron), the latter mesons (quark – anti-quark systems, e.g., pion, kaon).

Antimatter is identically composed of anti-quarks and anti-leptons, and these differ from the quarks and leptons of normal matter only by the opposite sign of charge. In the case of the electrically uncharged neutrinos, the situation is a bit more complicated. The question of whether anti-neutrinos must be distinguished from neutrinos, and thus, whether they are different particles, has not yet been experimentally resolved[3]. From the theoretical point of view both variants are conceivable but with different consequences for the Early Universe.

The three families of quarks and leptons represent exact copies of each other, with the only difference that, as indicated in Fig. 4.1, the masses increase from left to right. Thus, all particles that contain building blocks from

---

[3] The neutrinoless double β-decay, which has not yet been observed, will be able to answer this question.

families 2 and 3, are inherently unstable and decay back to the family 1. Why Nature affords this gimmick of triple manifold is one of the biggest open question of elementary particle physics. Are there perhaps further families? — Up to now, all experimental results exclude this.

The question about the deeper sense of the 3 families is all the more confusing because the "normal", stable matter surrounding us is composed only of the building blocks of the first family of quarks and leptons, and apparently in a completely satisfying way. Proton and neutron have (in shortened form) the combinations:

$$\text{proton:} \quad p = \quad (uud) \quad \text{charge} = +1$$
$$\text{neutron:} \quad n = \quad (udd) \quad \text{charge} = 0$$

A proton and an electron then combine to form a hydrogen atom. Neutrinos seem to be superfluous, all the more so, since they hardly interact — thus, they are moving "on the edge of existence". However, we will see later that they could possibly play a crucial role in the emergence of Life.

4.2   The interactions

The forces between the particles are mediated by so-called field particles. This concept is a bit more difficult to grasp intellectually and shall be made clear by an example shown in Fig. 4.2:

Two electrically charged spheres (particles) change their direction of motion due to the repulsive field (without meeting directly). The transferred momentum (force impact) is mediated by a photon, which is emitted by one sphere and re-absorbed by the other due to the change of the electric field while being in motion. The process is dynamic and occurs continuously during this motion process, which is expressed in the mathematical description by an integral.

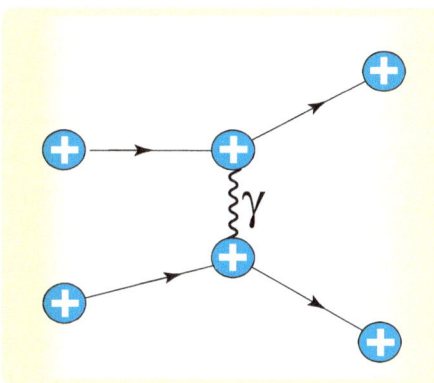

Fig. 4.2: Motion sequence of two electrically (here positively) charged spheres (particles). The transferred momentum (force impact) is mediated by a photon (denoted here as γ), which is emitted by one sphere and absorbed by the other, and thus causes a change of direction for both spheres. Depending on the signs of the charges ($++$, $--$ or $+-$, $-+$) the force is either repulsive or attractive.

In the «*strong*» interaction a total of 8 such field particles — called gluons — appear with the additional and non-trivial property that these (in contrast to the uncharged photon) also possess color charges themselves, which then entails a mathematical description of the interaction processes that is extraordinarily complex and computationally extremely costly. The mathematical model derived from all of this is called Quantum Chromodynamics (*chroma*, Greek: color) or simply QCD in reference to the classical color scheme.

Finally, the «*weak*» interaction is mediated by 3 field particles, W^+, Z^0 and W^-, which in contrast to the massless photons and gluons, have extremely large, and meanwhile also precisely measured masses, as these are $M(Z^0) = (91{,}187.6 \pm 2.1)\,\text{MeV}$ and $M(W^\pm) = (80{,}401 \pm 43)\,\text{MeV}$. This corresponds roughly to 97 times and 86 times the mass of the proton, respectively. These particles were predicted theoretically in the course of the unification of «*electromagnetic*» and «*weak*» interactions at the end of the 1960s and finally found experimentally at CERN in 1983[4]. The «*weak*» interaction is responsible, for example, for nuclear β-decay, which accounts for most of the radioactivity occurring in Nature. It also appears whenever neutrinos are involved in a reaction. Two examples may explain this, one is the β-decay of the neutron and the other is the elastic scattering of a neutrino by the deuteron. The latter is, for instance, of importance for the

---

[4] For the theoretical work on unification Sheldon Glashow, Abdus Salam, and Steven Weinberg were awarded the Nobel Prize in 1979, and for the discovery of the W and the Z^0 particle as the mediators of the «*weak*» interaction, Carlo Rubbia and Simon van der Meer received the Nobel Prize in 1984.

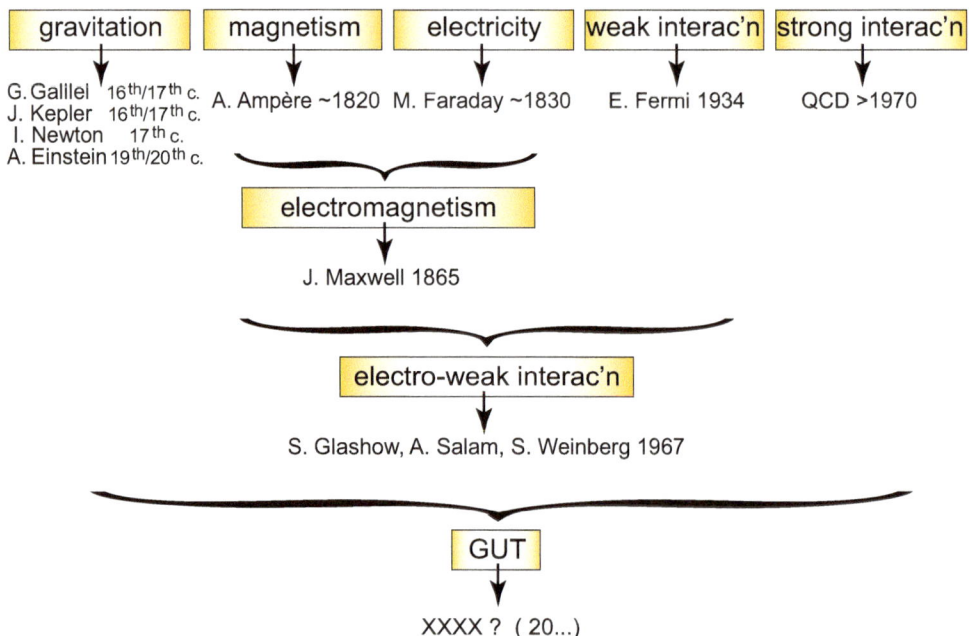

Fig. 4.3: The various interactions in physics and the attempt to unify them in the framework of a "Grand Unified Theory" (GUT).

detection of neutrinos from the Sun (we will not go into this detail here):

decay of a neutron: $\qquad n \quad \longrightarrow \quad p + e^- + \overline{\nu_e}$ $\quad$ W-exchange
$\qquad\qquad\qquad\qquad\qquad\qquad\qquad\qquad\qquad\qquad\qquad$ half-life $609.6 \pm 0.6\,\mathrm{s}$

neutrino reaction: $\quad \nu + d \quad \longrightarrow \quad \nu + d$ $\qquad\qquad$ Z^0-exchange

The situation of the different interactions is still extremely unsatisfactory. It is not conceivable that the Universe was equipped with a handful of interactions already at the earliest stage of its formation, but rather that there was initially only a single interaction, which, in the course of cooling, broke down to the interactions one knows today — roughly comparable to a phase transition from gaseous to liquid to solid. To find and formulate such a "Grand Unified" interaction is one of the great challenges for the future. That this can succeed has been impressively demonstrated in the past, e.g., by the unification of the «*electric*» and «*magnetic*» interactions into the «*electromagnetic*» interaction, expressed by Maxwell's field equations

(1865), or by the unification of the «*electromagnetic*» and «*weak*» interactions into the «*electroweak*» interaction, which followed 100 years later by the theory of Weinberg, Salam, and Glashow (1967) (see Fig. 4.3). This theory describes with impressive accuracy all experimental results on the «*electroweak*» interaction without exception, from the simple electronic circuit, to radioactivity, to supernova explosions.

It is also instructive to get a picture of the strength of each interaction. If one arbitrarily normalizes the strength of the «*strong*» interaction to "1", then the approximate relations listed in Tab. 4.1 result.

Tab. 4.1: Properties of the known interactions

interaction (IA)	rel. strength	range
strong IA	1	extremely short, $\approx 10^{-15}\,\mathrm{m}$
electromagnetic IA	1/137	infinite
weak IA	10^{-6}	extremely short, $\approx 10^{-17}\,\mathrm{m}$
gravitation	10^{-38}	infinite

Surprisingly, the interaction that we experience every day in our own bodies, namely gravity, is the weakest of all. It is completely negligible for processes which take place on a microscopic scale (e.g., for atoms and atomic nuclei). Conversely, the «*strong*» and «*weak*» interactions are macroscopically imperceptible due to their miniscule range. The infinite range of «*electromagnetic*» and «*gravitational*» interaction is here associated with the mathematical formula, according to which the interaction potentials die down only linearly with 1/distance.

Chapter 5

Beginning and End of Time and Space

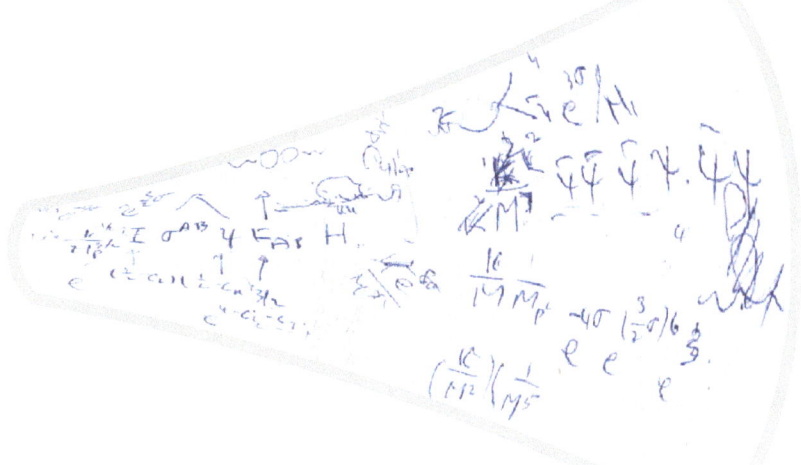

5.1 The epochs of the Universe

The Universe begins at the time $t = 0$ with the "Big Bang". Is this correct...?
does this make sense...? — There is no immediate answer to that, because
regrettably the state of the Universe eludes a consideration at $t = 0$ and,
of course, also its properties before this time. However, processes occurring
after the so-called Planck time of about 5×10^{-44} seconds seem to follow
consistently the laws of known physics. The present post-"Big Bang" theory
is therefore entirely conceived from well-established principles of physics. As
such, it is capable of describing all the multi-faceted signals stemming from
the earliest epochs with astonishing precision, so that a holistic picture of
these earliest processes in the Universe emerges. There is no alternative

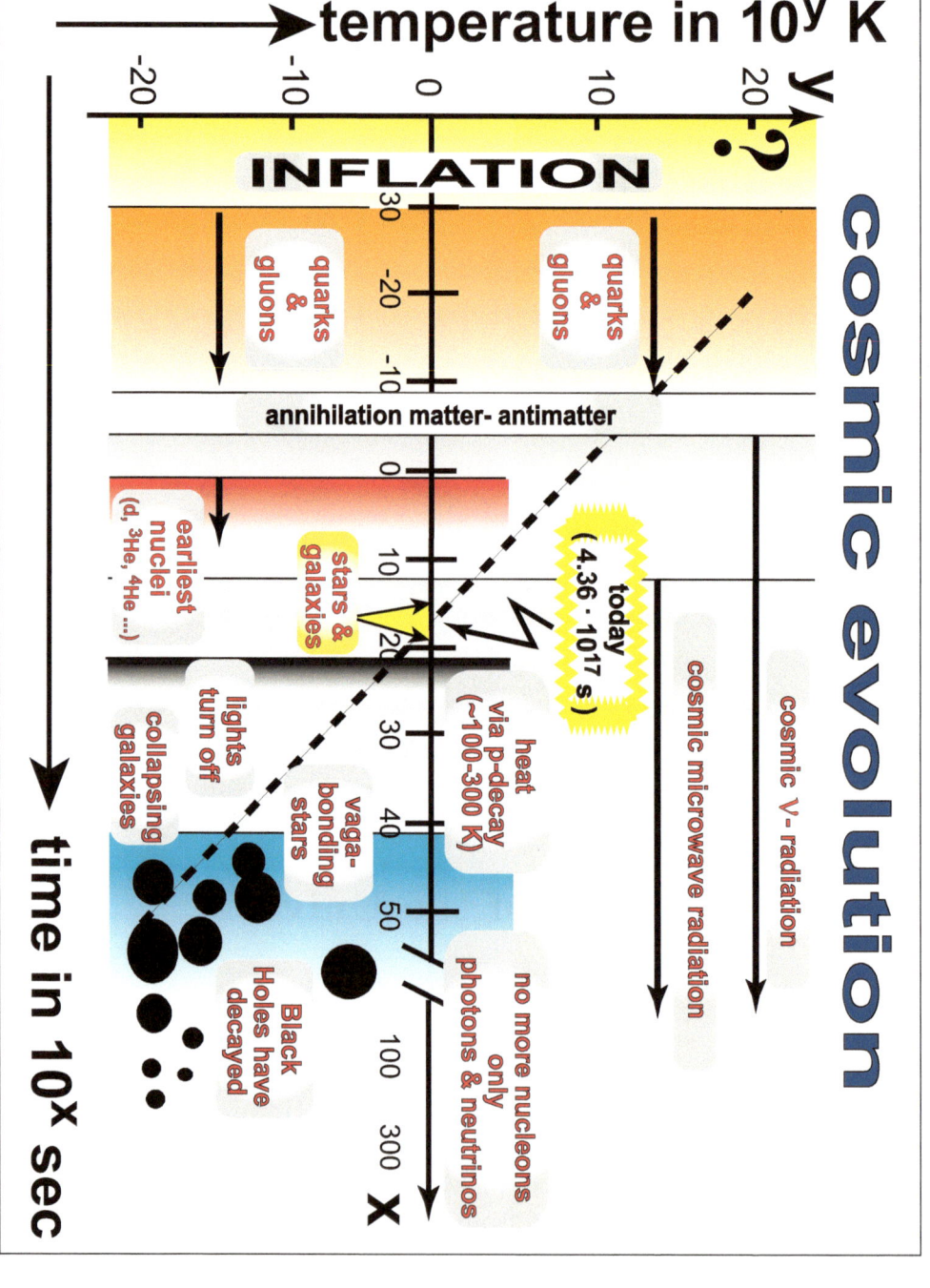

Fig. 5.1: Sequences of the cosmic epochs of the Universe

model in sight that would be anywhere near as powerful and effective as the "Big Bang" model.

In the "Big Bang"-concept, the Universe starts from a nearly point-like "structure" equipped with only energy and some universal interaction. It subsequently expands "explosively" (for lack of another term) and goes through different phases. However, with regards to the initial conditions, i.e., the magnitude of this energy, the kind of the interaction, the form of the "structure", or even more generally about the very cause of the Universe, no reliable statements in the language of physics can be made at present.

In Fig. 5.1 these individual evolutionary phases of the Universe are depicted. The plot in double-logarithmic form is somewhat unusual and needs additional explanation. Here, the time epochs span a range of nearly 400 !! orders of magnitude (powers of 10), starting at about 10^{-40} seconds to beyond 10^{300} seconds. For orientation, the era in which we live today is marked in the diagram. It indicates that at the present time the Universe is just 4.36×10^{17} seconds (13.81 ± 0.04 billion years) old. The temperature of the Universe is plotted on the ordinate from about 10^{-20} K to 10^{+20} K (see footnote [1]). The dashed line shows how the Universe cools over time due to its expansion. The temperature at the present time is 2.73 K (about $-270°$ C), which is close to zero ($\log 2.73 \simeq 0.44$) in this logarithmic plot. The temperature is determined by the cosmic heat radiation (or cosmic microwave background radiation), which floods the entire Universe. It has been measured over the past 30 years through a series of balloon and satellite experiments with ever increasing accuracy, currently up to the 6th decimal point [2].

First, some brief introductory remarks about the individual epochs; each of these will be examined again in detail in the chapters which follow.

The inflationary era following the Planck time

When making a back calculation in time for the cosmic evolution, one encounters at the so-called Planck time ($\approx 5 \times 10^{-44}$ s) a hard knowledge

---

[1] As a reminder, the Kelvin temperature scale follows the Celsius temperature scale, with only the zero point shifted: 0 K $= -273.15°$ C.

[2] From this temperature, the age of the Universe can be deduced by back calculation.

barrier. Classical physics, the theory of relativity, quantum mechanics, all of these almost irrefutable cornerstones of physics fail on all fronts at this point. A quantum gravity theory could perhaps come to the rescue, but such a theory does not yet exist. Thus, the question then arises as to how such a theory could be tested experimentally. A major sticking point is that in this era classical concepts such as time and space, and thus cause and effect, lose their meaning, leaving it unclear how the substitutes for these terms should eventually look like.

Only after the Planck time do the evolutionary processes become gradually more transparent, so that the sequence of events from there on can be understood on the basis of established laws of physics. At about 10^{-35} seconds a spectacular phase transition takes place in the Universe. During the period from $10^{-35} - 10^{-30}$ seconds the Universe expands by about $40 - 70$ orders of magnitude. The experimental evidence for such an inflationary phase transition is quite compelling and the classical equations of motion also allow for such a process. Fundamentally new physics must not be invented for this.

But how can this monumental inflation of the Universe be reconciled with the constancy of the speed of light? In fact, during this epoch only space and space-time expand at every given point, and thereby initially causally connected areas suddenly find themselves separated from each other. The term "causally connected" means that during this inflationary phase all information about the initial state is carried over into the expanding space and diluted everywhere in the same way. An exchange of information with super-luminal velocity does not take place to regions outside an "event horizon", which is given by the distance light can travel in 10^{-30} s. This also means that "our Universe" would be embedded in a much larger "Mega-Universe" with essentially equal properties everywhere. This has, of course, consequences as we will see later.

The first hundred nanoseconds

With the creation of space and time the Universe imprints an interaction into space-time. The elementary particles known to us, such as quarks, gluons, photons, leptons and possibly also extremely heavy and so far unknown "X" and "Y" particles [3)] are being created. Temperature and particle density remain high, so that all particles are in thermodynamic equilibrium, i.e., creation and decay are always in balance due to the intense interaction. Quarks and gluons move freely in a so-called quark-gluon plasma. To reproduce and study such a state is presently regarded as one of the most important research tasks, foremost at the CERN research center in Geneva.

After about 100 ns the expanding Universe has cooled down far enough so that the quarks and the gluons condense to the known baryons and mesons. To the baryons belong, for instance, protons and neutrons but also their antiparticles like anti-protons and anti-neutrons. To the mesons belong, for instance, the pions and kaons, which are the lightest particles in this group. The leptons, like electrons, muons, tauons with their related neutrinos, such as electron-, muon-, and tau-neutrinos, are subject to another, the «*weak*» interaction. They cannot make bound systems, and therefore remain free for all times.

The next process, which is now about to begin, is dramatic and could have already meant the end of the Universe — and that after less than a microsecond.

Matter-antimatter annihilation

After the freeze-out of the quarks and gluons into bound states like protons and neutrons, the way back is no more possible. But since matter-particles and antimatter-particles are created in the same way, the Universe should now be filled with exactly 50% of each of these two species. So far so good — however, these particles and antiparticles, as they are in close contact to each other, annihilate instantaneously, i.e., shortly after their creation (this in contrast to the free quarks and anti-quarks). In doing so, they eventually bring their mass-equivalent energy ($E = mc^2$) as photons directly back

---

[3)] This sounds like something "is thrown into the hat" for no reason. In fact, the initial interaction is not known, but there are theoretical models that can describe the initial baryo- and leptogenesis for the case that such "X" and "Y" particles once existed — theoretically quite compelling, experimentally not verified, though. But this comes with a testable prediction, namely proton and neutron must decay, however, with an extremely long half-life.

into the heat bath of the Universe. The process would cease soon and nothing would remain, except light! The Universe would be at the end of its evolution, and this after about a tenth of a microsecond.

Fortunately, Nature, at its early stage has built a tiny asymmetry into the interaction, which now leads to the very fortunate situation that out of about 10^9 matter particles, one is left over with no antimatter partner. The number 10^9 is an experimentally measured number and follows directly from the intensity of the cosmic microwave background radiation [4] as well as from the fact that no significant clusters of antimatter have ever been detected in the entire Universe. And what weighs even more heavily: not even a single heavy anti-atomic nucleus, such as an anti-carbon nucleus, for which no plausible creation mechanism out of normal matter exists, has been detected so far, despite intensive searches. That again says, this tiny remaining fraction of matter after this early annihilation is now that part, which the Universe of today is made of. The Universe and us have been very lucky indeed!!

The background neutrinos

From now on we speak of a (our) Universe, in which there is almost no matter left anymore, and in which the number of photons is more than 10^9 times larger than the number of nucleons (in this case protons and neutrons).

Of course, the matter-antimatter annihilation captures electrons and positrons (the antiparticles of the electrons) in the same way, so that their original number will also be reduced to about the same remaining fraction. However, electrons and positrons "survive" the annihilation phase much longer than nucleons and anti-nucleons, since the energy and intensity of the background radiation (and thus temperature and density) is still large enough for a comparatively long time (about 26 s) to reverse the annihilation process again and thus maintain an equilibrium between production and annihilation. In a shortened reaction equation (leaving out any intermediate steps) this can be expressed as follows:

$$e^+ + e^- \leftrightarrows \gamma + \gamma,$$

---

[4] Experimentally one finds a photon/baryon(nucleon) ratio of $(1.64 \pm 0.01) \times 10^9$.

where the double arrow indicates that in thermodynamic equilibrium the back and forth reactions are equally likely.

For neutrinos, the picture is completely different. Due to their purely «*weak*» interaction, the Universe becomes transparent already after about 100 ms, i.e., their mean free path between two interactions gets to be larger than the size of the Universe at this time. Thus, with no interaction in sight, they leave the thermodynamic equilibrium, in which all other particles so far remain. One says: "they decouple". Since neutrinos and anti-neutrinos do not annihilate, it follows that at the time of decoupling there ought to be about as many neutrinos and anti-neutrinos as photons in the Universe, namely about 10^{10} times as many as nucleons. That would be — as for the photons — still the case up to the present time.

As the Universe keeps expanding, the "neutrino gas" thins and cools down like a normal gas. Today's temperature of the "neutrino gas" (or what one may call "neutrino background radiation") can be calculated accurately to be $1.95\,K$, which corresponds to an average neutrino energy of about $160\,\mu eV$. So much the theory!! — because an experimental proof for these background neutrinos to exist is still lacking. And unfortunately, no viable proposal as to how an experiment for the detection of these cosmic background neutrinos could be carried out has so far been put forward. The difficulties lie, on the one hand, in the extremely low interaction rate and, on the other hand, in the extremely low energy that one would have to record from such an interaction event.

Yet, if the scenario described above were to be confirmed experimentally at some time in the future, it would be one of the most grandiose confirmations of the "Big Bang" model — and not only that, these neutrinos could provide a means to look back at the phase less than a second after the "Big Bang".

Nucleosynthesis in the Early Universe

We are now in a phase (at a fraction of a second) where the Universe contains only neutrinos, anti-neutrinos, photons, electrons and positrons, as well as a vanishing fraction of protons and neutrons. The Universe keeps cooling down at a rapid rate.

To build elements, protons and neutrons must fuse first. However, the early attempts to build a deuteron from a proton and a neutron fail over and over again. The photons of the thermal background radiation immediately photo-dissociate any such created deuteron owing to their enormous number advantage ($\gamma + d \longrightarrow p + n$). In fact, one has to wait until the Universe has cooled down to about 75 keV ($\sim 10^9$ K), which takes about 3 minutes. A race against time begins, because at this point the neutron is also partly "decoupled" and thus free. A free neutron though, is unstable and decays into a proton with a half-life of about 10 min, which means there is not a lot of time left for any fusion process to build heavy elements, as all of these require a deuteron at the initial stage. In this race, the rapidly expanding Universe just barely manages to synthesize ^{4}He via a chain of known nuclear reactions, each starting with the deuteron, then it is over. Temperature and density simply get to be too low. At the end of this phase (i.e., after less than 10 min), the Universe consists of about 75% protons and about 25% ^{4}He (plus tiny traces of deuterons (d), ^{3}He, ^{6}Li, and ^{7}Li). These mass fractions remain virtually unchanged to this day. And here again the "Big Bang" model shows its prodigious predictive power because the calculations of these primordial element abundances are confirmed with an amazing accuracy by experimental data on primordial stellar systems (e.g., globular clusters). Even the Sun, where the conversion from hydrogen to helium has been going on for 4.5 billion years (at a rate of about 610 million tons per second[5]), shows a largely unchanged primordial 75/25 hydrogen-to-helium ratio[6].

The mysterious background radiation

After the primordial nucleosynthesis, a period of quietness sets in. The Universe now consists predominantly of thermal photons with a tiny fraction of protons, ^{4}He nuclei, electrons, and of course the non-interacting and already decoupled neutrinos. This state persists for about 380,000 years. If one could look at the Universe from the outside, it would appear as a sphere of light (similar to a sun), first in the spectrum of X-rays, which then changes

---

[5] This rate can be trivially determined from the solar radiation power of 1.367 kW/m^2 arriving on Earth, as will be shown later.

[6] The value of the hydrogen/helium mass ratio in the Sun is about 68% to 31%.

with increasing size into the spectral range of ultraviolet. It would not be possible to look into the sphere because the photons are constantly scattered by the free electrons in the inner area and thereby continuously change their direction of motion. Also inside the sphere one would see only diffuse light, without being able to make out any specific source. Any attempt to bind an electron to a proton to make a hydrogen atom, similar to the case of the deuteron, also fails over and over again because the background radiation ionizes the hydrogen atom immediately after its formation. In a shortened reaction equation this can be expressed again as follows (note "$h\nu$" is used for photons at atomic energy scales):

$$\mathrm{H} \; + \; h\nu \; \leftrightarrows \; p \; + \; e^{-}$$

After about 380,000 years, the temperature of the Universe has dropped to the point that the electron can successfully be captured by a proton to make a hydrogen atom. The surrounding radiation is now no longer able to ionize it. The photons cannot interact anymore, they decouple and move from now on in a straight line, which in turn means: the Universe gets transparent. This will also be the last time that the thermal photons interact with the rest of the Universe.

With the expansion continuing, the Universe cools according to the known laws of thermodynamics, so that today's precisely measured "thermal microwave background radiation" corresponds to a mean temperature of 2.725 K.

Structure formation

Relatively early on, i.e., already after about 100 million years, the first stars and proto-galaxies are formed, which are initially in an extremely turbulent phase of evolution before they finally reach a stage that is similar to that of today. Unfortunately, little is known about the details of the physical processes that cause this clumping during this phase, except that this must have been a purely gravitational effect. It is, however, likely that so-called "dark matter particles" play a decisive role here. For these there are two prime candidates on the list, the axion and the WIMP.

The axion particle, named by the Nobel laureate Frank Wilczek after a laundry detergent on the American market [7], has the advantage that it unavoidably comes into play by a simple, yet effective physics concept of a symmetry property of the «*strong*» interaction. This symmetry makes the «*strong*» interaction not to distinguish between matter and antimatter — this in strong contrast to the «*electromagnetic*» interaction. — And not only that !! — The axion particle possesses just by chance the right properties for the structure formation in the cosmos. So, it solves two problems from two different areas of physics at the same time.

According to the theory, the axion has the following properties,

- it is electrically neutral,
- it is an extremely light object $(10-100\,\mu\text{eV})$ (thereby perhaps even much lighter than a neutrino),
- it is generated "cold", and is already "decoupled" from birth, i.e., it does not take the temperature of the Universe at any time; thus it can "fall into" the smallest gravitational potentials and initiate structure formation very early on,
- it is generated in an immense number (about $10^{90}-10^{100}$) in the earliest stage of the Universe and still floods the Universe today,
- it is subject to all known interactions («*gravitational*», «*strong*», «*weak*» and «*electromagnetic*»), however, the rate of interaction is so low that it may only be gravitationally detected as dark matter,
- it has exotic quantum numbers which may help to devise experimental detection methods to directly tune in to this property.

The other particle on the list of candidates is called the **W**eakly-**I**nteracting-**M**assive-**P**article or WIMP. It is theoretically far less compelling and requires some additional ad-hoc assumptions, each of which depends on the theoretical model. It is an exclusively gravitationally and weakly interacting particle and is somewhat similar to the well-known neutrino. It is also a heavy particle, but how heavy is not quantifiable theoretically. Reasonable estimates put its mass between that of an iron atom and that of a lead atom. If true, the WIMP would also be an effective structure-forming particle,

---

[7] There was a good reason for this because the axion "washes away a problem": Without axion, the «*strong*» interaction would distinguish between matter and antimatter. Matter and antimatter are, however, found to experience an identical interaction strength.

Fig. 5.2:   A galaxy in possible size comparison to its outer halo.

and elaborate computer simulations indeed show that when such an exotic particle is included in these calculations, the filamentary structures of large galaxy assemblies, as they develop after an elapsed time of 13.8 billion years, turn out to be remarkably similar to those observed today.

This particle also solves a second problem. Due to its large particle mass, it moves relatively slowly at all times and thus can be gravitationally captured by galaxies[8]. In fact, there is experimental evidence from our Milky Way, and similarly from almost all other galaxies, that these are each surrounded by a widely extended, merely gravitationally interacting (and thus invisible) halo (not teapots[9]), whose combined mass comes to about 10 to 50 times the stellar mass in a galaxy (see Fig. 5.2). The WIMP particles possess the properties to produce exactly these halos — but as it turns out, also axion particles may have this property.

---

[8] The escape velocity in a typical galaxy is about $300 - 400$ km/s.
[9] Cf. section 2.2 → statement 5.

Whatever the cause may be, the first emerging structures act as the seedlings for the formation of stars and galaxies. They set the gravitational contractions in motion, which lead to the first proto-galaxies. Each time, when large collections of hydrogen contract and a critical pressure and density is reached, hydrogen-to-helium fusion ignites and a star is born. Initially these processes are quite frequent and violent and only during the next $1-2$ billion years a certain calming down with a largely constant birth rate sets in.

The processes of star births still take place today and can be observed again and again in quite some detail in every galaxy including in our own Milky Way Galaxy. The birth rate of stars whose sizes are similar to that of the Sun is about $2-3$ stars per year for the Milky Way. One can now make a rough estimate by extrapolating this rate to the approximately 50 billion galaxies in the observable Universe to get a "universal" star birth rate of about 5000 stars per second!! Since the far distant and thereby youngest galaxies were much more active, a more realistic estimate of the star formation rate may even be closer to twice this number.

Supernovae and the making of elements

Unfortunately, so far the Universe still hasn't managed to synthesize elements heavier than helium. This is going to change in the phase we are now approaching, a few billion years after the "Big Bang". One may recall that stars about the size of the Sun have a typical lifetime of about 10 billion years. During this time they quietly and unobtrusively fuse hydrogen into helium, before they simply extinguish at the end after a brief burst. Stars with masses beyond about $10-100$ solar masses go through a life cycle $10-1000$ times shorter and then perish in a gigantic explosion that tears the entire star apart. These explosions are so powerful that they outshine an entire galaxy for a few days (see Fig. 5.3). This also means they are visible with today's space telescopes to the most distant corners of the Universe. In such a "supernova" explosion all the elements in the periodic table up to the heaviest ones known to us, like uranium as well as all the trans-uraniums (e.g., neptunium, plutonium, etc.) are synthesized within a few seconds. Interestingly, as the Fig. 5.1 shows, we are still in the midst of this phase today, i.e., 4.36×10^{17} seconds after the "Big Bang".

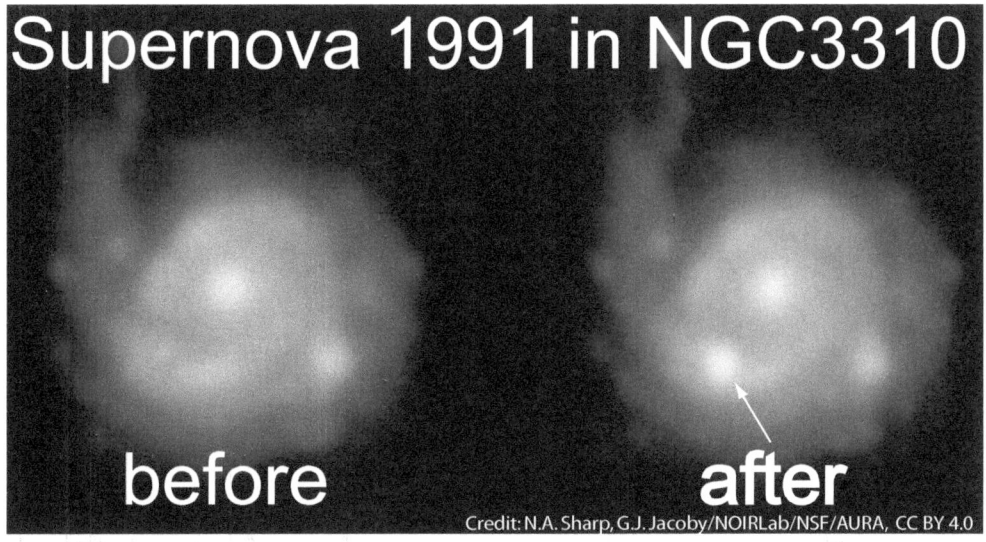

Fig. 5.3:   Supernova 1991N in the NGC3310 galaxy, distance about 47 million light-years.

The physical processes, which take place during a supernova explosion are reasonably well known, and the absolute luminosity of those events are also well determined. Supernovae (here in particular the ones of type Ia) can thus serve as "standard candles" to determine distances and from these to infer the Hubble constant, which describes how fast the Universe expands with increasing distance.

Supernovae are rare events. In the Milky Way their frequency is about 2 supernovae per 100 years. This translates to about 200 – 400 million of such supernova events that may have already occurred in the Milky Way during its existence, whereby one already assumes that in the earliest times the star birth rate was quite a bit higher than today.

In the end, supernovae are the cosmic cauldrons, where all elements that are needed for the making of Life as well as those needed for sustaining it, are produced. This also implies that we owe our existence to such a supernova explosion to the extent that from its remnants the Sun and finally the solar system with planet Earth were formed 4.6 billion years ago.

The lights go out

The amount of hydrogen in the Universe is abundant, although not infinite. The Sun will remain in the hydrogen-burning phase for about another 4.5 billion years, thereby gradually inflating to a Red Giant and finally perishing as a White Dwarf. This inevitable evolutionary process will already affect Earth in about 500 – 900 million years from now, when the Sun slowly gets hotter and the increase in temperature eventually makes life on Earth impossible. At the end of its development the Sun will have swallowed the inner planets Mercury, Venus and Earth before finally going as a White Dwarf into the cosmic grave in about 5 – 6 billion years from now.

But the Universe is not finished yet. The amount of hydrogen is still sufficient to keep star evolution active for about $10^{20}-10^{21}$ seconds. That is roughly 1000 times the present age of the Universe. After that, the Universe slowly (very slowly!) transforms to become dark. Neutron stars and Black Holes are more and more formed at increasing numbers. Interestingly, neutron stars continue to radiate small amounts of energy, because even protons and neutrons, which so far have been assumed to be stable, eventually decay and inject the decay energy as heat in to the remaining star. The nucleon half-lives are likely in the order of $10^{34} - 10^{37}$ years (about $10^{41}-10^{44}$ seconds). Every theoretical model of particle physics, which seeks to extend the existing interactions towards Grand Unification, gets this extremely rare decay of the proton (and the neutron) as an unavoidable byproduct. The decay could be initiated by the hypothetical and earlier mentioned "X" and "Y" particles. Experiments on proton decay have therefore been ongoing for more than 20 years with no positive signal being observed so far. The current best lower bound for the half-life of the proton is given by a Japanese collaboration as 1.6×10^{34} years [10]. This corresponds to a single proton decay per year in about 360,000 tons of water.

The decays of the proton and the neutron feature several decay chains, where a positron and several photons or leptons are produced at the end. In a White Dwarf as well as in a neutron star almost the full mass-equivalent

---

[10] K. Abe et al., *Search for proton decay via $p \to e^+\pi^0$ and $p \to \mu^+\pi^0$ in 0.31 megaton-years exposure of the Super-Kamiokande water Cherenkov detector*, Physical Review D 95, 012004 (2017), see also same title same authors, arXiv:1610.03597v2 [hep-ex].

of the respective nucleon of about 938 MeV is converted into heat. This can still lead to a considerable and permanent heating to about 300 K (room temperature) in a neutron star once it has cooled down to this level.

The most likely decays of the proton and the neutron are:

$$p \longrightarrow e^+ + \pi^0 \qquad\qquad\qquad + \simeq 938\,\text{MeV} \rightarrow \text{heat}$$
$$\hookrightarrow \gamma + \gamma$$

$$n \longrightarrow e^+ + \pi^- \qquad\qquad\qquad + \simeq 800\,\text{MeV} \rightarrow \text{heat},$$
$$\hookrightarrow \mu^- + \overline{\nu}_\mu \qquad\qquad \text{loss due to escaping neutrinos}$$
$$\hookrightarrow e^- + \nu_\mu + \overline{\nu}_e$$

The process of nucleon decay, which slowly leads to the demise of the neutron star, will end after about 10^{45} seconds.

Black Holes do not escape the fate of decay either. They constantly lose energy, and thus mass, by the so-called Hawking radiation, though the more massive the Black Hole, the smaller its energy outflow. The lifetime of a Black Hole scales with the third power of its mass, and from Hawking radiation it can be calculated that a super-massive Black Hole takes about 10^{100} seconds to finally die explosively ($T_{1/2}\,[\text{s}] \simeq 8.4 \cdot 10^{-17} \cdot M^3\,[\text{kg}]$). This leaves the Universe after about 10^{100} seconds with only electrons, positrons (from nucleon decay), neutrinos, and photons [11]. After another roughly estimated 10^{300} seconds also the electrons and positrons find to each other due to their electric attraction. They annihilate to photons, so that at the very end only neutrinos and photons remain in a still expanding Universe. Neutrinos and photons were already present from the beginning in high abundance, so that in retrospect the star evolutions, the galaxy formations, the supernova explosions etc. represented only an entirely insignificant intermezzo.

Interesting is a final consideration, which results from quantum mechanics. An essential and also experimentally well confirmed property of the micro-world is the so-called Heisenberg uncertainty relation. It says, energy and time can never be determined exactly at the same instant. The uncertain-

---

[11] Note that the Universe is still electrically neutral.

ties of these two quantities are linked by the relation $\Delta E \cdot \Delta t \geq \hbar = \text{const.}$! Applied to the condition of the Universe, this implies that, as the energy density keeps decreasing, the energy uncertainty ΔE also decreases. In order to fulfill the above relation, the time uncertainty Δt must therefore increase in the same way. In return, one is led to conclude: "time slowly dissolves", it is no longer possible to determine the distance between two points A and B. Such a result is certainly not completely counter-intuitive, because how can one locate these two points in a Universe, in which "there isn't anything left anymore", and who is "one" after all?

QUESTIONS:

how does one want to know all this? – simply by forward projection using the known laws of physics.

could it be otherwise? – of course, there is still a lot of unknown physics.

what about gravitational waves? – they also represent background radiation but they won't change the arguments — although an experimentally verified theory of quantum gravity does not exist.

does that still have anything to do with Science and Physics ? – NO, no one will live to see it, this future scenario, as it is described, is not falsifiable.

BUT:

it is of course intellectually extremely inspiring and immensely stimulating to conceive and further develop future scenarios within the framework of existing and still deepening physical knowledge, even if these are not verifiable in their final instance.

Chapter 6

The Inflationary Universe

6.1 Unheeded hints, questions without answers

The essence of physics is to ask questions, and frequently the most profound questions appear to be trivial and banal.

Here is an outline of the supposedly most trivial questions. Quite remarkably, they are of cosmological relevance and are directly or indirectly related to the phenomenon of inflation.

D. Frekers, P. Biermann, *Universe, Neutrinos, Stars and Life*, https://doi.org/10.1007/978-3-662-70729-6_6

(1) why is the sum of angles in a   Carl Friedrich Gauss (1777–1855)
      triangle $\alpha + \beta + \gamma = 180°$?   (in German: Gauß)

(2) why is it dark at night?   Heinrich Wilhelm Olbers
                                                    (1758–1840)

(3) why do my arms lift up   Ernst Mach (1838–1918)
      when I turn?   Albert Einstein (1879–1955)

(4) why does the Foucault pendu-   Jean-Bernard-Léon Foucault
      lum rotate, how does it know   (1819–1868)
      its swing plane?

(5) why are there so many different
      things in the Universe ?
      why is the Universe nevertheless   unknown
      so terribly empty?
      what's the point of all this?

Re (1): **Carl Friedrich Gauss** is in the eyes of these authors one of
the greatest and most important scientists, mathematicians, physicists,
astronomers, geodesists of the whole last millennium. Students of natural
sciences know well the Gaussian function, the Gaussian distribution, the
Gaussian integral theorem, the Gaussian number plane and the gauss (G)
as the unit of the magnetic field strength, to name only a few of the
"Gauss" measures. Gauss was mainly concerned with geometry, and he
was apparently puzzled that Nature had chosen this completely unique
and flat Euclidean geometry from an infinite number of other possible
choices. Only in this special geometry, the sum of angles in a triangle
is 180°. In every positively curved (elliptic) geometry – similar to a sphere
in the 4-dimensional space – it is larger, and in every negatively curved
(hyperbolic) geometry – similar to a saddle in the 4-dimensional space – it is
smaller. This is illustrated in Fig. 6.1 for the case of a 2-dimensional surface
embedded in a 3-dimensional space. Gauss succeeded by an ingenious
experiment to measure very accurately the angles of the triangle formed by

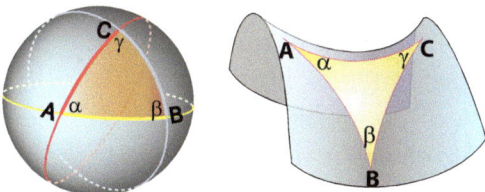

Fig. 6.1: Representation of curvature. Taking the shortest connecting distance between the points of the triangle A, B, C, then the sum of angles in the triangle on the left is greater than 180°, and on the right it is smaller.

Fig. 6.2: The German 10-DM bill was dedicated to Carl Friedrich Gauss and his scientific work. The bill (here only front side) entered circulation in 1991 and was replaced by the Euro on January-1, 2002.

three mountains, the "Brocken", the "Hoher Hagen" and the "Inselberg" [1] with sides 69 km, 85 km and 107 km. He found, probably to his great surprise, that within the accuracy of the measurement the sum of angles was indeed 180°. Of course, at that time he was not able to give an explanation why Nature preferred Euclidean geometry. It may sound strange, but cosmic inflation is the only theory that gives an elegant and conclusive explanation for this result as well as for the general flatness of the Universe.

The German 10-DM bill, which was only briefly in circulation before the Euro was introduced, was dedicated to the scientist Carl Friedrich Gauss on both the front and back side (Fig. 6.2).

Re ②: Until far into the 20th century the greater part of the general public was still convinced that the Universe was infinitely extended (the problem of infinity in physics has already been alluded to earlier in this book). However,

---

[1] These 3 mountain peaks are located within about 100 km from the German city of Göttingen, where at its university Gauss became the director of the Göttingen Observatory in 1807. The "Brocken" is located about 70 km north-east, the "Hoher Hagen" about 17 km south-west and the "Inselberg" about 100 km south-east of the city of Göttingen.

the fundamental problem of an infinite Universe was already formulated in 1823 by the German astronomer **Heinrich Wilhelm Olbers** and is known as Olbers paradox[2]. It is derived from the simple and seemingly trivial question *"why is it actually dark at night?"* Whoever hastily answers "because the Sun does not shine at night" did not grasp the profoundness of the question.

An infinite Universe naturally has an infinite number of stars. Although the apparent brightness of a star decreases with the square of the distance, the number of stars lying in the field of view increases with the square of the distance, so that the summed apparent brightness gets to be constant and infinite, both during day and night! The presence or absence of the Sun is irrelevant. There was not much effort by scientists to solve this paradox in the 19th century; only A. Einstein took up this problem again around 1917. He concluded that the Universe must be finite — but just how can one stabilize the outer edge and prevent the Universe from collapsing gravitationally? His idea of introducing a cosmological constant that would regulate everything also proved to be contradictory very early on.

Re ③: The question of the origin of the so-called apparent or inertial forces introduced in classical mechanics (i.e., the centrifugal force, the Coriolis force, the inertial force or acceleration force) was for a long time the subject of a lively correspondence between **Ernst Mach** and **Albert Einstein**. The problem can be formulated in a simple way as a question: *"why do my arms lift when I turn around?"* (or synonymously "what is the origin of the centrifugal force?"). A turn (or rotation) requires a reference system against which this rotation takes place. This reference system cannot be the Earth because it rotates around the Sun, which in turn rotates with the galaxy. Somehow everything rotates — but around what? In particular, what happens finally far away from Earth, perhaps even far away from the Milky Way, e.g., in a spaceship in "dark" intergalactic space, when no obvious reference frame can be made out anymore? Are the centrifugal forces the same as on Earth?

E. Mach was of the opinion that these apparent forces must emanate from

---

[2] H. W. M. Olbers, *Über die Durchsichtigkeit des Weltraums*, Essay of May 7, 1823, published in "Astronomisches Jahrbuch für das Jahr 1826", p. 110-121. Ed. J. Bode., Berlin.

the distant stars, because they are always and everywhere present and appear to rotate in exactly the opposite direction to a rotating observer. If true, then one must also assume that the stellar density is homogeneous and equal in all directions, which was much later found to be indeed the case, at least for scales beyond about 1 billion light-years. However, a conclusive mathematical concept for this assumption that takes into account the expansion of the Universe, has not yet been developed. We will see later that the phenomenon of inflation plays a decisive role here as well.

Re ④: An almost identical question arises from the movement of Foucault's pendulum. **Jean-Bernard-Léon Foucault** had presented this pendulum for the first time to an astonished Parisian public, in order to show with it that Earth actually rotates. The pendulum plane, or swing plane, remains stationary, and the Earth rotates so to speak, under this pendulum. At the North and South Poles this produces an apparent rotation of the pendulum of one full revolution in 24 hours, and in Paris, one full revolution in 31.8 hours [3], corresponding to its location at 48.8° latitude [4]. And again the question arises *"how does the pendulum know what its swing plane or reference frame is with respect to which it must be stationary?"* Once more, the concept of distant stars and their homogeneous distribution comes into play.

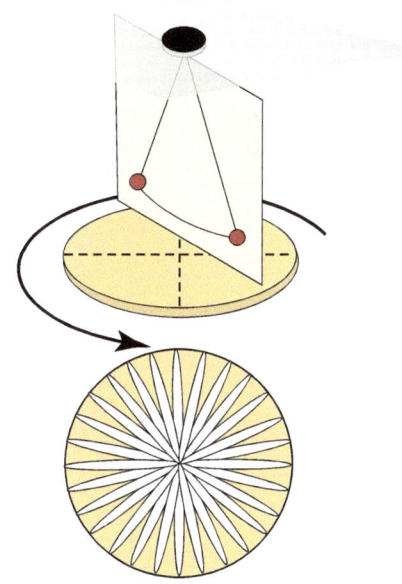

Fig. 6.3:   The Foucault pendulum. The Earth (indicated by the disk) rotates under the pendulum and produces in the course of one revolution the star-shaped figure as is schematically indicated here. The lines must not necessarily close.

---

[3] For the city of Münster (latitude 51.96°) the full rotation of the pendulum takes 30.47 hours.

[4] The rotational frequency is calculated as $\omega_{\text{Paris}} = \omega_0 \cdot \sin(48.8°)$ and one full rotation of the pendulum as $T_{\text{Paris}} = T_0 \cdot (1/\sin(48.8°))$.

The experimental situation is indicated in Fig. 6.3. The Foucault pendulum in the city of Münster with a suspension length of 28.75 meters has an oscillation period [5] of about 10.7 seconds and performs about 10,200 oscillations per revolution with an equal number of oscillation figures, as indicated in Fig. 6.3.

Re ⑤: Finally we come to the question ***"why are there so many different things in the Universe and yet, why is the Universe practically empty?"*** [6]? Here physics must give up, because there is no answer. A comparison universe that one could consult and explore is unfortunately not in sight. So at present, only a retreat to a philosophical level remains, using, for example, the « *Anthropic Principle* », which says « *the Universe is the way it is because we observe it* »; it depends on a multitude of seemingly independent variables, and these are remarkably selected to enable the development of intelligent life in a way that 13.81 billion years after its birth the Universe is capable of reflecting upon itself. Any slight changes to these variables would have likely caused this development to fail, and the initial question would not have come up at all.

6.2   Mass, time, space, space-time

Mass

To regard mass simply as a chunk of material is certainly a generally useful concept. However, the appearance, the color, the consistency or the hardness of the material, or even the very subjective notion of a collision with another mass is ultimately and solely a result of the « *electromagnetic* » interaction. All these different manifestations of substances and matter can be traced back without exception to the collective properties of the electrons in atomic bound states or to the distribution of the electric charges in a solid.

---

[5] The period of oscillation of an ideal (mathematical) pendulum is $t = 2\pi\sqrt{l/g}$ with l the length of the suspension and g the acceleration due to gravity ($9.81\,\mathrm{m/s^2}$).

[6] The average density of the Universe is $\approx 5 \times 10^{-30}\,\mathrm{g/cm^3}$ for "normal" matter or about one hydrogen atom per $\mathrm{m^3}$.

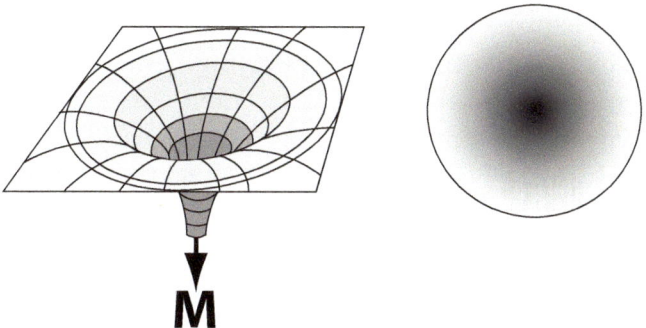

Fig. 6.4: Mass as a "dent in space" (here as a 2-dimensional dent in a 3-dimensional space). The projection is to represent the compression of the spatial structure in the presence of a mass.

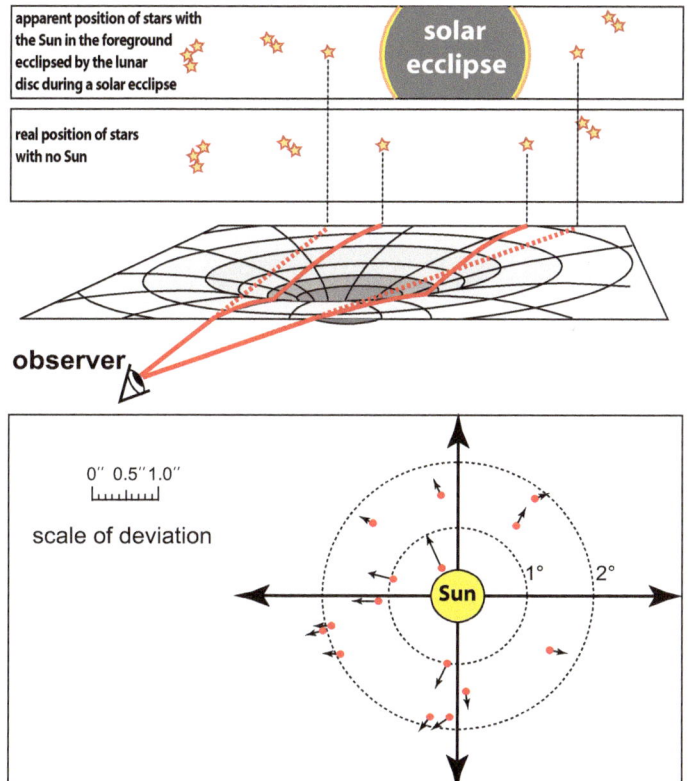

Fig. 6.5: Change of star positions near the solar disk measured during the solar eclipse on May 29, 1919. The upper image visualizes the effect, the lower shows the measurement results for stars in the 2° neighborhood around the solar disk. Despite the large solar mass, the curvature effects are small and lie in the range of one arcsecond (1″) and smaller (scale of arrows in the upper left). The fact that the arrows in the image do not point exactly radially outward is due to the limited resolution at that time.

All masses are also subject to the «*gravitational*» interaction, and in General Relativity mass has a more general meaning. Mass makes (or even is) a "dent" in the space-time structure. The "dent" resists a change of motion as is well-known from a massive object. This extension of the concept has an additional advantage: empty space is not necessarily at energy zero. In practice, this means that empty space can be "shifted up and down" on the energy scale. For the understanding of inflation this becomes an important aspect, and for an interpretation of the "dark energy" in the Universe this could also be an important clue. Figure 6.4 will clarify this, where a 2-dimensional surface is shown with a "dent" in the 3-dimensional space. The projection visualizes this effect more clearly. Mass creates a densification of space in the spatial sphere surrounding it.

This "dent" in space changes the geometry of space and creates curvature parallels are suddenly no longer parallel and the sum of angles in a triangle is no longer 180° (one may want to remember Carl Friedrich Gauss in the previous section). It was therefore one of the most fascinating discoveries in the 20th century, when Arthur Eddington was able to actually measure the curvature of space produced by the mass of the Sun during the course of a total solar eclipse on May 29, 1919. At that time there was an especially favorable constellation of the solar disk, as it overlapped with the center of the Hyades star cluster, so that the apparent position changes of many stars could be detected at the same time. The curvature of space produced by the Sun caused stars whose calculated positions were close to the solar disk to appear further away from the disk in their straight line of sight. As expected, the measured effects were small, amounting to only fractions of an arcsecond. This is shown in Fig. 6.5.

With today's technical abilities, these mass-related effects are, of course, confirmed to a high degree of accuracy and, as it appears, in ever-improving agreement with General Relativity.

Time

Time is still a very strange coordinate in physics. It knows only one direction — probably due to causality — but nevertheless, it is not absolute (see an artist design in Fig. 6.6). Together with space they both make up

what one calls space-time. This
is due to the fact that one of the
two coordinates, time or distance,
is redundant. The time span Δt
and the distance Δx are linked by
the natural constant c, the speed
of light, by virtue of the relation
$\Delta x = c \cdot \Delta t$. It follows that one can
express a distance by a time and
vice versa, e.g., 1 meter is equivalent
to ≈ 3.3 nanoseconds and one year is
equivalent to $\approx 10^{13}$ km (=1 light-
year).

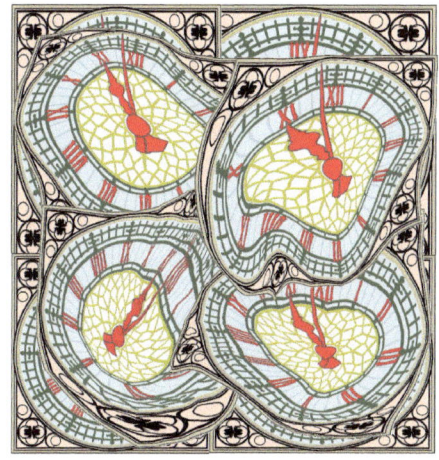

Fig. 6.6:   Warped time.

It gets to be utterly counterintui-
tive, when one has to realize that the speed of light is a finite and above all
absolute quantity. It is independent of the direction of motion of the system
emitting the light signal, or formulated differently, the particular velocity
v of a system and the velocity of light c do not add up to $c' = c \pm v$, as
one might assume. In the chapter about Relativity and Black Holes this
phenomenon is highlighted again in more detail.

Still, Johannes Kepler was convinced of the infinity of the speed of light
without recognizing the contradiction, because then all light of the Universe
would be instantaneously everywhere. Light could not be turned off.

Space, space-time

Since light is the carrier of information, or even more strictly formulated,
since any type of information can only be transported with the maximum
speed c, it follows immediately that each look into the distance is equivalent
to a look into the past. In daily life this hardly plays a role because all times
are too short to worry about the few nanoseconds of time delay when looking
at your conversational partner. For the Sun, the distance of about 8 minutes
is already considerable and for the binary star α-Centauri, the 4.3 years is
substantial. Today's event on this neighboring star will therefore not enter

the event horizon "Earth" until about 4.3 years from now.

The situation is explained in Fig. 6.7. The systems **A** and **B** are stationary. The cones around them describe the region, which becomes visible for each of these systems after a time t. Let system **B** be about 163,000 light-years ($x = c \cdot t$) away from system **A**. At a time $t =$ Feb-24-1987, **A** (by now having arrived in time at **A'**) observes how **B** explodes as a supernova[7]. This means that 163,000 years after the event in **B**, this same event enters the event horizon (time cone) of **A**. Also **B** has arrived in **B'** in the meantime, saying that a new event in **B'** will enter the event horizon of **A** only after yet another 163,000 years.

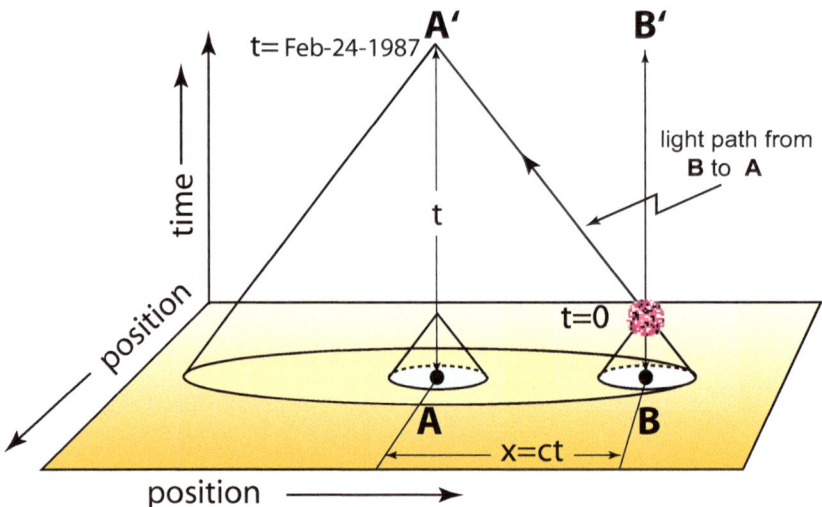

Fig. 6.7: Event horizons: At time $t = 0$ a supernova explodes in **B**. If **B** is 163,000 light-years away from **A**, **A** will become aware of this event only 163,000 years later, i.e., when arriving at **A'** on the time axis.

A passage through time

Knowing that a look into the distance is equivalent to a look into the past, it is instructive to go on a time travel into the past (see Fig. 6.8).

---

[7] On February 24, 1987, Supernova SN1987 A exploded in the Large Magellanic Cloud, about 163,000 light-years from Earth.

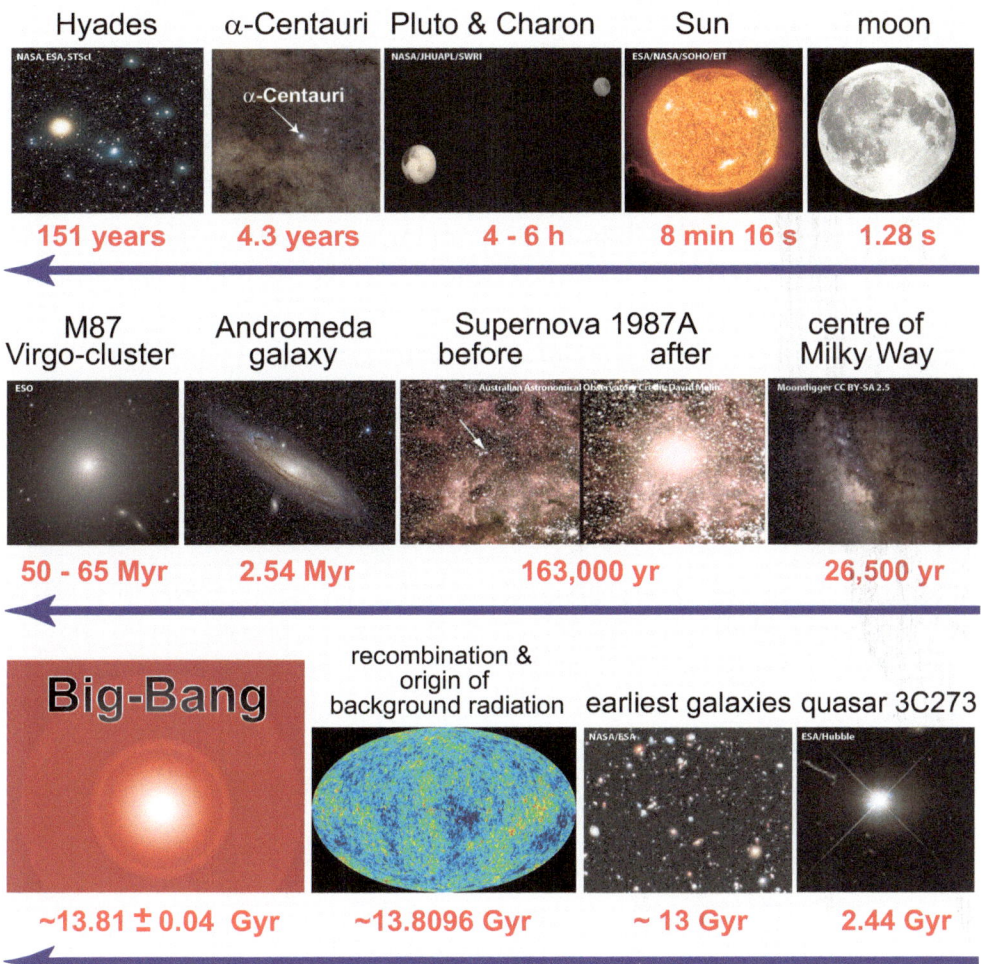

Fig. 6.8: Time travel back in time to the "Big Bang".

The moon is 384,400 km away from Earth, that corresponds to 1.28 light-seconds, and the Sun is already 8 minutes and 16 seconds away. The "no-more planet" Pluto with its companion Charon already brings it to about $4-6$ hours depending on position, and the next solar system, the double star system α-Centauri A and B, is 4.3 light-years away.

The aforementioned Hyades star cluster, consisting of a loose cluster of about 350 closely spaced stars, is about 151 light-years away, and the distance to the center of the Milky Way with its supermassive Black Hole is a safe 27,000 light-years, give or take a few hundred. The closest galaxy to us is the Andromeda Galaxy, and we see it today as it appeared 2.54 million years ago. The Virgo cluster with its approximately 2000 galaxies is between $50-65$ million light-years away, and some of its inhabitants are probably watching the extinction of the dinosaurs on Earth.

A big leap takes us 2.44 billion light-years into the past to the closest quasar, 3C271 (Quasi-stellar Radio Source), whose distance from Earth increases by about 50,000 km per second. Quasars emit enormous energies, estimated to be up to 10^{12} times that of the Sun. These are generated when huge amounts of matter are swept into an extremely massive Black Hole.

The first proto-galaxies discovered in recent years date back to about 13 billion years ago and may be the first compact objects after the "Big Bang". Their escaping velocities are about 98% of the speed of light.

Finally, we reach the time of recombination, i.e., about 380,000 years after the "Big Bang" or about 13.8096 billion years ago. It constitutes the visible limit of the Universe, and its evidence is the microwave background radiation, which has now cooled to 2.725 K. Yet, the far more interesting physics is still hidden behind this curtain, and only neutrinos and gravitational waves have penetrated it unhindered. And up to the endpoint "Big Bang" it is still a long and truly exciting way to go.

6.3   Static or dynamic Universe?

The introduction of the theory of relativity and especially its extension to the theory of General Relativity in the early 20th century triggered an extraordinary dynamic in the two branches of science, i.e., cosmology and astronomy, but at the same time also produced numerous aberrations and erroneous concepts. Even Albert Einstein may have stumbled over his own equations because the physical consequences seemed to be so unrealistic and absurd that a firmly established world order suddenly seemed to be up for debate. At that time, though, experimental evidence for one or the other variant of a cosmo-theory did not exist, with the sole exception of the Michelson-Morley experiment, which testified to the constancy of the speed of light.

Fig. 6.9: The pioneers of early cosmology

Still in 1917 Einstein was of the opinion that the Universe was finite and static but his so-called field equations did not provide this property. The addition of a cosmological constant "Λ" by hand was supposed to solve this problem, a great stupidity ("größte Eselei") as he admitted later; because in a finite and static universe that is literally held together "from the outside" by a (single) cosmological constant, any movement and thus any change of the gravitational conditions would inevitably lead to a collapse.

Also the de-Sitter model[8], which was discussed until the 1930s, represented a particularly simple solution of the field equations, but it had a serious flaw — this universe was absolutely empty, no masses, no forces, either dynamically expanding or static, depending on the choice of parameters.

The Russian physicist and mathematician Alexander Friedmann[9] had already in 1922 worked out a largely unnoticed proposal for a solution that did not necessitate a cosmological constant and only assumed a homogeneous mass distribution in the Universe[10]. The equations of motion derived from the field equations gave either an expanding or contracting Universe, depending on the mass density.

Georges Lemaître[11] developed the same equations in 1927 independently

---

[8] Named after the Dutch cosmologist Willem de Sitter (1872 – 1934).

[9] Alexander Friedmann (1888 – 1925), Russian physicist and mathematician.

[10] A. Friedmann, *Über die Krümmung des Raumes*, Zeitschrift für Physik 10, 377 (1922).

[11] Georges Lemaître (1894 – 1966), Belgian physicist, priest and later papal prelate.

of A. Friedmann, but these appeared in a simpler notation [12]. Moreover, he showed that any dynamical solution of the field equations, when calculated back to an initial time $t = 0$, leads to a singular, point-like Universe (the "Big Bang"). Lemaître is therefore also considered the «*father of the Big Bang*». Furthermore, in 2018, the Hubble law named after E. Hubble was extended by his name and is now called the Hubble-Lemaître law.

These pioneers of the early cosmology are shown in Fig. 6.9.

A short mathematical side trip:

Einstein's field equations in relativistic formulation consist of a system of 256 coupled 2nd order nonlinear differential equations in 4-dimensional notation. They can be broken down to a set of easily manageable equations by exploiting symmetries and assuming isotropy and homogeneity of mass distributions and mass densities. Friedmann's equations of motion are such equations; they imply quite reasonable assumptions and are written in their simplest form without a cosmological constant:

$$\text{Eq. (I)} \quad 2\frac{\ddot{R}}{R} + \frac{\dot{R}^2 + kc^2}{R^2} \;=\; -8\pi G\frac{p}{c^2} \quad \left\{ = 0 \;\; \text{für } p = 0 \right\}$$

$$\text{Eq. (II)} \qquad \frac{\dot{R}^2 + kc^2}{R^2} \;=\; \frac{8\pi}{3}G\rho(t)$$

Meaning of the parameters:

R size of some arbitrary scale (e.g., Mpc)
$\dot{R}$ time derivative of this scale (velocity)
$\dot{R}/R$ relative time derivative of this scale (identical to the Hubble constant)
         (e.g., (km/s)/Mpc = unit of the Hubble constant)
$\ddot{R}/R$ relative acceleration of this scale
         (positive for acceleration, negative for deceleration)
k parameter of curvature
         ($k = -1, 0$ or $+1$) for negative, zero, or positive curvature

---

[12] G. Lemaître, *Un Univers homogène de masse constante et de rayon croissant rendant compte de la vitesse radiale des nébuleuses extra-galactiques*, Annales de la Société Scientifique de Bruxelles A47, 49 (1927).

c speed of light
G gravitational constant
$\rho(t)$ mass density at time t (implies all forms of mass and energy)
p pressure in the Universe –
               for a radiation dominated Universe p is positive
               for a cold-matter-dominated Universe $p = 0$.

In the case of a cosmological constant Λ (e.g., dark energy) replace:

$$\text{Eq. (III)} \qquad \rho \to \rho - \frac{\Lambda c^2}{8\pi G}$$

$$\text{Eq. (IV)} \qquad p \to p + \frac{\Lambda c^4}{8\pi G}$$

The first Friedmann equation Eq. (I) says: The strength of the deceleration depends quadratically on the speed of the expansion, with a possible pressure term either increasing the deceleration, if $p > 0$, or leading to an acceleration, if $p < 0$ (e.g., for a corresponding $\Lambda < 0$ in Eq. (IV)). The second Friedmann equation Eq. (II) states: The square of the relative expansion velocity is compensated by the product of gravity and mass density — or: With decreasing mass density (i.e., with increasing age of the Universe) the relative expansion between any two objects slows down (note: Λ in Eq. (III) contributes to the mass density). To better understand these relationships, simply put $k = 0$ in Eqs. (I) and (II).

In retrospect — and surprisingly — it turns out that the two Friedmann equations (Eq. (I) and Eq. (II)) can also be derived from classical Newtonian physics. Only the curvature parameter k and the constant Λ, if both are not 0, remain in these equations as relics of General Relativity.

Here an intuitive explanation of the above equations to avoid confusion when reading the following chapters.

Assume there are two cars A and B on a highway (local frame) driving at some high speed (expanding Universe). A drives at a maximum speed allowed. B follows A at some distance but moves at a lower speed than A. Thereby the distance between A and B grows with time, or in other words, A and B get further and further apart, each of them sees the other one moving away (Hubble-Lemaître law). As time goes on, B manages to slowly catch up to the speed of A (i.e., to the maximum speed), and as a consequence, the increase of the distance between A and B slowly comes to a halt (Hubble constant gets smaller with time). At time infinity, A and B finally drive at the same speed, the separation between them is now fixed (Hubble constant gets to be zero at time infinity). Of course, the same happens to all cars on the highway. (In mathematical terms; $H_0 \propto 1/t,\ \ t \to \infty \Rightarrow H_0 \to 0$.)

6.4   A dynamic Universe — it expands

The question whether the Universe is indeed expanding, as suggested by Friedmann's equations, was finally answered experimentally with "yes, it does" by the American astronomer Edwin Hubble in 1929. This discovery was epochal, and immediately memories of the statement "tamensi movetur" (lat.) or "eppur si muove" (it.) ("and it moves nonetheless") come back, which Galilei is said to have made, according to legend, almost exactly 300 years earlier, after he was forced to renounce the heliocentric world view.

In a long and painstaking work, Hubble correlated the spectral frequency shifts with the distances to 24 galaxies, and he discovered that with increasing distance the redshifts of the incoming light invariably increased[13]. Of course, it was known at that time that redshifts, i.e., the stretching of wavelengths λ, occur when the source moves away from the observer in direct analogy to the Doppler effect in acoustics (see Fig. 6.10).

---

[13] Edwin Hubble, *A relation between distance and radial velocities among extragalactic nebulae*, Proceedings of the National Academy of Sciences, Vol. 15, No. 3, 168 (1929).

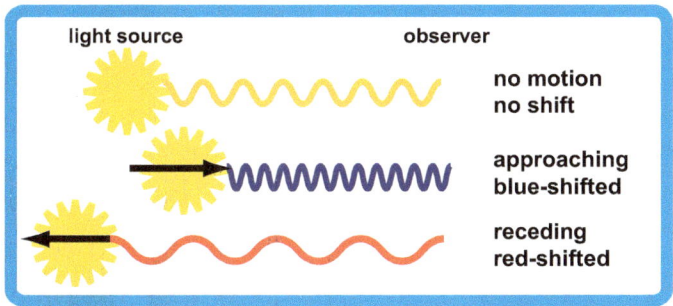

Fig. 6.10:   Optical Doppler effect

The redshift z is for small z ($z \lesssim 0.2$) in a simple way related to the velocity v:

$$z = \frac{\lambda_{\text{observed}} - \lambda_{\text{source}}}{\lambda_{\text{source}}} = \frac{\Delta\lambda}{\lambda_{\text{source}}}$$
$$v = z \cdot c .$$

For the distance determinations, E. Hubble could use the period-luminosity relation of Cepheids that Henrietta Leavitt [14] had already discovered in 1912 [15]. Thus, by observing a number of Cepheids in these 24 galaxies, he was able to determine their respective distances out to 2 Mpc. Thus, the Hubble-Lemaître law for the escape velocities of galaxies, as is valid until today, reads:

$$v = H \cdot D$$

with H the Hubble constant and D the distance.

---

[14] Henrietta Swan Leavitt (1866 – 1921), American astronomer. In 1912 she discovered the period-luminosity relation of variable stars (called Cepheids) in our companion dwarf galaxy, the Small Magellanic Cloud. The period-luminosity relation establishes a relationship between the absolute luminosity of Cepheids and their variable periods — the greater the amount of light emitted the longer the period. Establishing this relationship was possible because the distance to the Small Magellanic Cloud [about 202,000 light-years, value of 2013, see "The Astrophysical Journal", Vol. 230, No. 1, 59 (2013)] was already known and thus assumed approximately the same for all Cepheids it contained. Henrietta Leavitt's work laid the foundation for the use of the Cepheids as "standard candles" for distance determinations of galaxies up to about 100 million light-years (about 30 Mpc). One may note that her work was barely recognized at the time.

[15] Henrietta S. Leavitt and Edward C. Pickering, *Periods of 25 variable stars in the Small Magellanic Cloud*, Harvard College Observatory Circular 173, 1 (1912).

Still given by E. Hubble as $465 \pm 50 \, \mathrm{(km/s)/Mpc}$, today's much more exact value of the Hubble constant is $H = 74.3 \pm 2 \, \mathrm{(km/s)/Mpc}$. It has been determined by many independent methods, although the scientific community may still argue about the size of the errors and the slightly lower value of $H = 67.7 \pm 0.4 \, \mathrm{(km/s)/Mpc}$ determined by the Planck collaboration [16].

Question: If all stars and galaxies move away from the reference point Earth, is the Earth then the center of the Universe after all?

This question is easily answered by a small computer experiment. One generates on a transparency an arbitrary starry sky with the Earth as center, represented in the following Fig. 6.11-**A**. Each point is now transformed according to the Hubble-Lemaître law — the further away, the greater the flight velocity and the greater the distance to Earth after a certain elapsed time. The result is recorded on a second transparency, here in Fig. 6.11-**B**. If one places both transparencies on top of each other with the Earth as reference point, the image **C** results. If one chooses any other reference point (indicated by the green arrow in the image **B**), the result is image **D**.

Any other observer in **B** sees qualitatively the same image as the one on Earth — all stars and galaxies seem to be moving away. In both cases the subjective impression is left as if one were at the center of the Universe.

One can go even one step further, and impress a rotation on the Universe, which then produces picture **E**. Stars and galaxies move not only radially away from the observer, but now also tangentially. Experimentally, though, a rotational motion of galaxies in our Universe has not yet been detected.

---

[16] Planck Collaboration, *Planck 2018 results. VI. Cosmological parameters*, Astronomy & Astrophysics 641, A6 (2020), und arXiv:1807.06209v2 [astro-ph.CO] (2019).

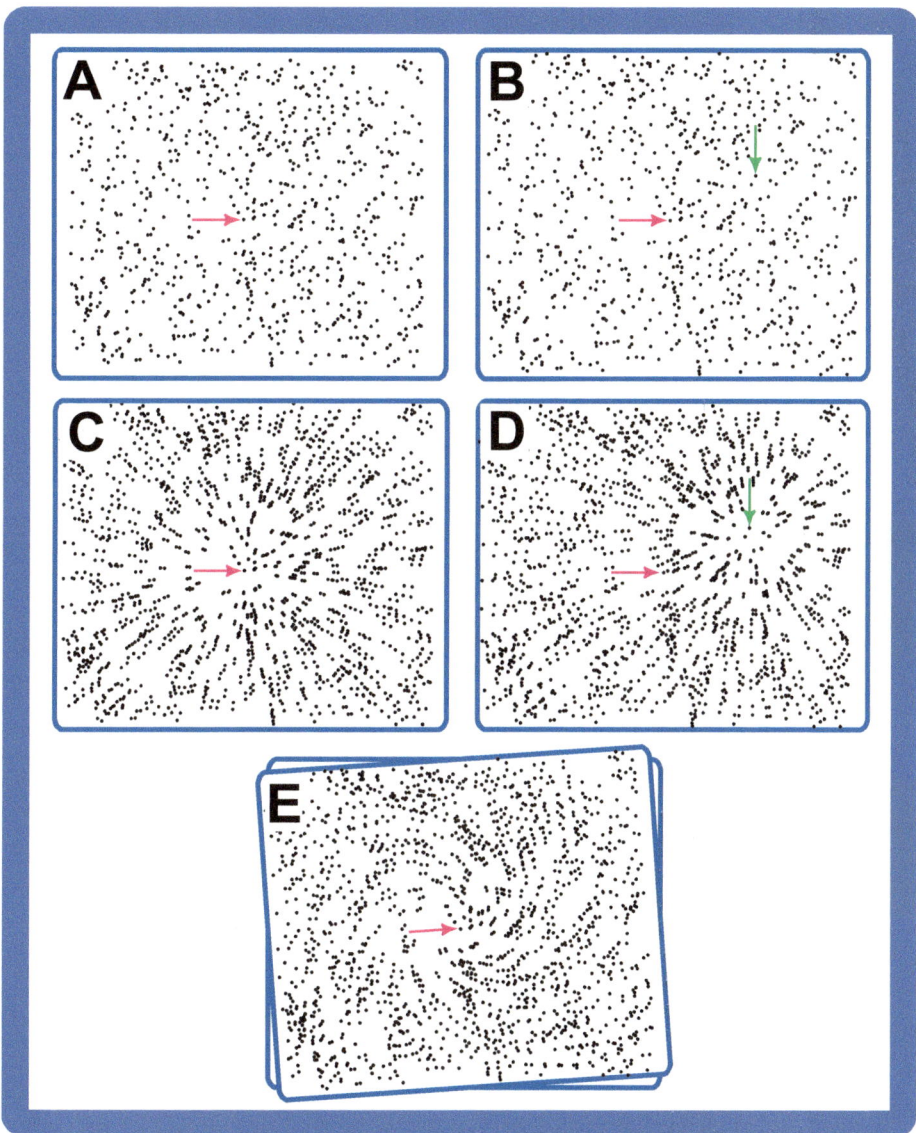

Fig. 6.11:
A: The Earth (red arrow) as the apparent center of the Universe with the starry sky.
B: The starry sky expanding according to the Hubble-Lemaître law with the Earth as center.
C: The superposition of **A** and **B** shows how the stars move radially away from the Earth.
D: The superposition of **A** and **B** with respect to any other star (green arrow) shows how stars move radially away from that star in the same way.
E: The superposition of **A** and **B** with respect to the Earth (red arrow) and with rotation of the Universe (here by 4°) shows how stars now move both radially and tangentially away from the Earth.

6.5   The problem of the "back calculation"

The realization that the Universe is expanding also means that distances between galaxies were smaller in the past. From the same argument follows that the temperature of the background radiation must have been higher in former times, meaning the radiation was more energetic. The solutions of Friedmann's equations give this in a conclusive and unmistakable way. But even if one discards the fact that these equations lead to a nonphysical singularity at the time $t = 0$, this logic has already some issues from the very beginning. One of them is the so-called «*horizon problem*», another one is the «*flatness problem*». It is conceivable that Einstein already reflected on these problems because both of them do not occur in his static and finite Universe.

The horizon problem

The horizon problem can be explained in a simple way by means of a thought experiment:

We assume the Universe is 14 billion years old and the Earth is in contact with the galaxy "Happyface" 😄 , which is 1 billion light-years away. In a static Universe, Earth and 😄 can exchange information 14 times (7 times to, 7 times back).

In an expanding Universe we may first go back to the epoch with redshift $z = 1100$ (which is the time of recombination). At this time the size of the Universe is a factor 1100 smaller than today, i.e., Earth and 😄 are now only 0.9 million light-years apart. However, the Universe at this time is just 380,000 years old [17]. Earth and 😄 have so far not been able to exchange any information. Earth must first wait — as shown in Fig. 6.7 — until 😄 enters its event horizon. Since Earth and 😄 are moving apart during the waiting time, this will finally be the case after about 2.5 million years. Because this reasoning also applies to every other point in today's Universe, it implies that today's observable Universe can be decomposed into about 1.3 million (1100^3) sub-areas that were not causally connected at that time.

---

[17] The relation between radius R of the Universe, age t of the Universe and the redshift z is given in good approximation by: $\left(\dfrac{t}{t_{today}}\right)^{2/3} = \dfrac{R}{R_{today}} = \dfrac{1}{(1+z)}$

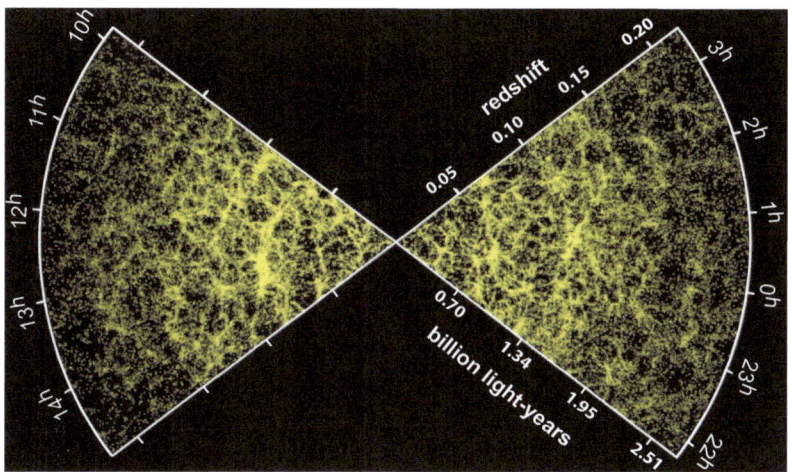

Fig. 6.12:   Galaxy distribution measured with the Australian "2dF" (2 degree Field) telescope. The redshifts of about 66,000 galaxies were determined within a measurement time of 5 years. The hour circles (right ascensions) of their positions are also given.

This example shows that as a Universe ages and expands, more and more galaxies enter the event horizon "Earth". If we now look at the spatial distribution of galaxies as in Fig. 6.12, we see that the newly added galaxies have the same pattern, the same distribution, the same density as those already within the event horizon. The question that follows is:

> How can it be that the galaxy distribution and galaxy density outside our (and any other) event horizon are exactly the same as they are inside the respective event horizon?

One may want to remember the Foucault pendulum. If Mach's principle is valid, which says that the swing plane remains stationary because of the distant stars, then this already gave an indication for the extraordinary homogeneity of the Universe on large scales. This homogeneity is obviously not broken by the newly added areas continuously coming into the event horizon.

A similar reasoning arises from the microwave background radiation that was released in the redshift epoch at $z \simeq 1100$. Photons of this radiation coming from the right and from the left and those coming the from the front and from the back, meet "today" in the detector for the first time. They originate from causally non-connected areas as it seems, and yet indicate

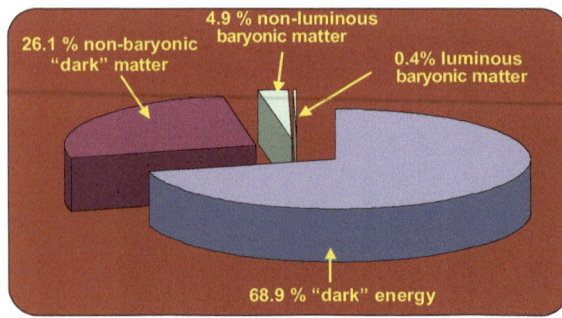

Fig. 6.13: Components of the Universe. The term "luminous" means here: luminous in the visible optical region. (Refer to chapter 8 and page 114 for more information including specific error estimates.)

exactly the same temperatures in these areas. This immediately raises the question:

> How can it be that temperatures were exactly the same in areas that apparently never had any causal contact in the past?

The flatness problem

The Universe shows no curvature on large scales, it is flat with a Euclidean geometry. One may recall Carl Friedrich Gauss, who was already puzzled by this finding on Earth. This phenomenon is all the more noteworthy because there are density variations of up to 44 orders of magnitude on local scales. For instance, the average density of the Universe is about 5×10^{-30} g/cm^3, and in neutron stars it is about 10^{14} g/cm^3. The Universe is homogeneous but at the same time extremely clumpy.

Another astonishing phenomenon adds to it. The matter density (better: energy density) ρ of the Universe determined by many different measurements is equal to the critical density ρ_{crit} within the experimental error. The critical density ρ_{crit} is the sum of gravitational energy (negative sign) and kinetic expansion energy (positive sign), and it is characterized by the fact that the observable expansion of the Universe would come to a halt at time $t = \infty$ [18]. This is expressed by a formula:

$$\left(\frac{\rho}{\rho_{\text{crit}}}\right) = \Omega_{\text{crit}}$$
$$= \Omega_{\text{B}} + \Omega_{\text{dark}-\text{M}} + \Omega_{\text{dark}-\text{E}}$$
$$= 0.99 \pm 0.01 \quad (\text{expmtl.})$$

---

[18] For deriving this relation one uses in physics the virial theorem.

Here Ω_B is the density of the "normal" or baryonic matter, Ω_{dark-M} that of the unknown, only gravitationally interacting "dark" matter and Ω_{dark-E} the part of the completely strange "dark" energy. The latter is equivalent to a cosmological constant "Λ" — and yes — discarded by Einstein, it comes back a century later, but in a different physics garment. The respective experimentally determined values for each of these are about 5%, 26%, and 69% [19] (see also chapter 8). The fractions are illustrated in Fig. 6.13.

The fact that this experimental Ω-value is compatible with $\Omega = 1$ is perhaps not very surprising. What is surprising, though, is the ensuing realization that in an expanding, i.e., our Universe, this value must be tuned to $\Omega = 1$ with extreme precision in the earliest stages to keep this Universe stable. Already a deviation in the order of "$\varepsilon = 0.000 \cdots 40 - 70$ zeros !! $\cdots x$" would have catastrophic effects. A value greater than 1 would lead to the early collapse of the Universe, and a value smaller than 1 would lead to a too rapid expansion, making the formation of structures such as galaxies and stars impossible. This reasoning follows directly from Friedmann's equations, but it is also intuitively plausible and understandable. First the mathematical argument:

The following relationship can easily be extracted from Friedmann's equations:

$$[1 - \Omega^{-1}(t)] \cdot \rho(t) \cdot R^2(t) = k \quad [k = 1, 0, -1],$$

where $R(t)$ is again a scale factor to be defined arbitrarily. If $\rho(t) = \rho_{crit}$, then $\Omega(t) = \Omega_{crit} = 1$ and the left expression is zero for all finite scales. Also for the curvature k it follows $k = 0$.

Since $\rho(t) \propto R^{-3}(t)$ (as density = mass per volume), then from the above equation it follows :

$$\rho(t)R^2(t) \propto 1/R(t),$$

so that

$$[1 - \Omega^{-1}(t)] \cdot R^{-1}(t) = k \quad [k = 1, 0, -1].$$

---

[19] The luminous (visible) stellar fraction contained in Ω_B is only 0.4%. Background neutrinos and the background photons contribute only a negligible part to the total energy density.

If after $t = 13.81$ billion years (i.e., today) one finds that

$$\Omega(t = \text{today}) = 0.99$$

then, in order to satisfy the above equation (with $k = -1$):

$$R^{-1}(t = \text{today}) = 99 \text{ [arb. units]}.$$

During the Planck epoch, however, the scale $R(t)$ is 40 to 70 orders of magnitude smaller, and thus $R^{-1}(t)$ in the above equation is 40 to 70 orders of magnitude larger, so that at this time we have:

$$\Omega^{-1} = 1 + 0.0000..... \approx 40 - 70 \text{ zeros} !!.....000x$$

or

$$\Omega = 0.99999..... \approx 40 - 70 \text{ nines} !!.....9999x,$$

which requires a considerable fine-tuning for the initial conditions of the Universe!

An intuitive explanation:

In an expanding, non-static, homogeneous Universe, the event horizon continuously expands with time, and new domains enter this event horizon at the speed of light. Since the critical density is calculated from the gravitational energy and the kinetic energy of the expansion, a steady increase or decrease of a given mass density leads to an ever increasing imbalance between gravitation (attraction) and expansion (escape), which causes the Universe to collapse in the first case or to expand further and faster in the second.

And again the question:

Who or what can tune the initial conditions for the development of the Universe in such an extremely precise way — and why should She or He or It do this ?

Inflation as the solution

Cosmic inflation is an idea proposed by Alan Guth [20] to circumvent the problem of fine-tuning the initial conditions [21]. The idea was subsequently further developed by Andrei Linde [22] and is now a widely accepted theory.

Inflation describes a phase transition of the Universe, which is set somewhat arbitrarily at about $10^{-35} - 10^{-30}$ seconds in order to keep it, for reasons of simplicity, out of reach of the unknown and causality-violating physics in the immediate vicinity of the Planck time of $t_P \approx 5 \times 10^{-44}$ s and the Planck temperature of $T_P \approx 10^{32}$ K. As a consequence of this phase transition, the Universe (or its space-time) enlarges by about $40 - 70$ orders of magnitude, depending on the choice of parameters. Space-time regions, which so far had been causally connected and had shared common state parameters, are now getting separated and all the original information about the initial state are carried over into the expanding space. The exchange of information among areas, which from now on lie beyond the range of light, is thus interrupted. Only at later times, and even up to 13.81 billion years, do these event horizons touch again (see e.g., Fig. 6.7), whereby the areas that got separated developed over time in exactly the same way, at least as far as temperature, density and structure formation is concerned. Assuming there was also homogeneity at the time of causality, then this property is, as well, fully retained in all areas. Inflation therefore turns out to be an elegant model capable of solving the horizon problem and the flatness problem at the same time.

The situation is graphically illustrated in Fig. 6.14. Causally connected areas during the pre-inflationary era (here: below the red line) get separated as a result of inflation and meet again at some later point in time.

The physical processes that generate inflation have not yet been described in

---

[20] Alan H. Guth, Professor for Theoretical Physics and Cosmology at Massachusetts Institute of Technology (MIT), USA.

[21] Alan H. Guth, *A possible solution to the horizon and flatness problems*, Physical Review D 23, 347 (1981).

[22] Andrei D. Linde, Professor for Physics at Stanford University, USA.

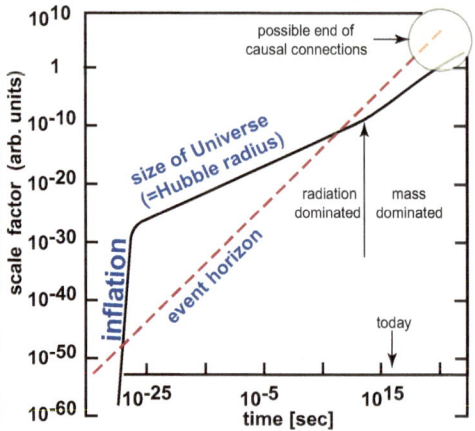

Fig. 6.14: Expansion of the Universe and the evolution of an event horizon as a function of time (over more than 50 !! decades). The "Hubble radius" is here synonymous to the size of the Universe; it increases according to the Hubble-Lemaître law. The expansion shows a kink at about 2 million years ($\approx 6 \times 10^{13}$ s). Up to this time, mainly radiation contributed to the energy density of the Universe, and only after cooling to about 2000 K is the energy density generated via masses. This raises the expansion velocity slightly[23]. —Starting at about 10^{20} seconds, there could be a transition into regions that were not causally connected before inflation. What happens then will not be known until this case occurs.

a uniform way from the theoretical side. The theories still have speculative components. Of course, this may also be attributed to the fact that, although there is ample evidence for an inflationary expansion of the Universe, the experimental data are still insufficient to differentiate sharply enough between different models. Nevertheless, one should not be mistaken: the current theories start from known physics principles and do not bring in crude fantasies that are merely tailored to inflation. Moreover, useful theoretical models are naturally expected to be able to make experimentally testable predictions. As an example, the occurrence of micro-structures in the cosmic background radiation is a common prediction of inflation theories, which can be readily tested experimentally. These micro-structures are, for one, characterized by particularly fine-grained fluctuations in the temperature spectrum, but even more decisively by tiny swirls of the micro-

---

[23] The energy density ε is defined as energy per volume, i.e., $\varepsilon = E/V \propto E/R^3$. Since the wavelength of electromagnetic radiation increases with expansion ($\lambda_R = \lambda_0 \cdot R$) and $E_{em} = \hbar/(2\pi\lambda_R)$, it follows for the energy density of radiation $\varepsilon \propto 1/R^4$, while for the matter energy density $\varepsilon \propto 1/R^3$ holds. In simple models, this results in the matter-dominated Universe to expand faster than the radiation-dominated Universe.

wave polarizations. These patterns were imprinted into the radiation by gravitational waves during the short inflationary era, as an echo of the "Big Bang", so to speak.

In 2014, the BICEP2 collaboration published experimental data interpreted exactly along these lines [24], though this interpretation was retracted a year later because ordinary cosmic dust could also have caused such a polarization effect [25]. Needless to say, these experiments are continuing, with attempts now being made to include spectral regions of the background radiation, where such perturbations are negligible. Just to give an idea here of the nature of the expected structures, the temperature spectrum measured by the BICEP2 collaboration is shown in Fig. 6.15 with the polarization vectors of the radiation plotted for a selected angular range [24]. However, caution is advised not to jump to any premature conclusions.

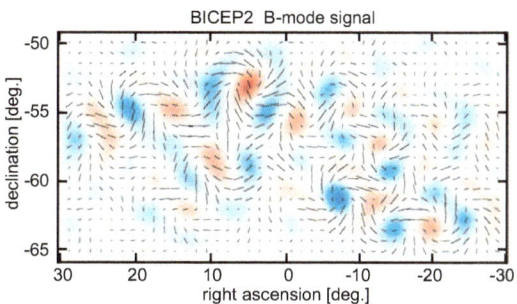

Fig. 6.15: Microstructures measured by BICEP2 in the temperature and polarization spectrum of the cosmic microwave background radiation over the indicated region of right ascension and declination [24]. The colors red and blue denote a temperature difference of $\delta T \approx 10^{-6}$ K at an average temperature of the radiation of 2.725 K. The spectrum with the polarization pattern imprinted is consistent with the theoretical prediction of an echo from the inflationary phase. However, it can also be explained trivially as a consequence of an interaction of the radiation with cosmic dust.

Unquestionably, more experimental information is needed to come to a deeper understanding of the dynamics of cosmic inflation and perhaps to bring to light the actual cause that triggered it. In the category of relevant

---

[24] P. A. R. Ade, et al., *Detection of B-Mode Polarization at Degree Angular Scales by BICEP2*, Physical Review Letters 112, 241101 (2014) and same authors, same title, arXiv:1403.3985 [astro-ph.CO] (2014).

[25] W. N. Colley, J. R. Gott, *Genus topology and cross-correlation of BICEP2 and Planck 353 GHz B-modes: further evidence favouring gravity wave detection*, Monthly Notices of the Royal Astro-nomical Society 447, 2034 (2015) - and J. R. Gott, W. N. Colley, *Reanalysis of the BICEP2, Keck and Planck Data: No Evidence for Gravitational Radiation*, arXiv:1707.06755 [astro-ph.CO] (2017).

experiments, special importance is given to background neutrinos. The signatures of inflation imprinted into the spectrum of the background neutrinos have likely been preserved in a much more distinctive and original form than in the spectrum of the background photons, as one may recall: neutrinos decouple already 100 ms after the "Big Bang" and thus possess a much more vivid memory of the inflationary phase than the background photons, which made their way through space 380,000 years later. However, the various ideas for detecting cosmic background neutrinos that have been put forward so far have all proved to be impracticable in the end. The experimental difficulties are simply unsurmountable for now.

Again something in mathematical language:

The description of the physics during the Planck era is incomplete and unknown in detail, simply because during this phase all current physics models violate the causality principle. This is easy to understand: The so-called Planck length is tied to the de-Broglie wavelength [26] $\lambda_{dB} = 2\pi\hbar/(mv) = 2\pi\hbar c/(m\beta c^2) \approx 2\pi\hbar c/(mc^2)$ $[\beta = v/c]$ of an object (i.e., a particle, or synonymously an energy). In the Planck era this wavelength becomes larger than the event horizon, which — in anticipation of a later chapter on Black Holes — is given by 2π times the horizon radius of the mass (= circumference of the event horizon). Or in short: A wave-mechanical (quantum-mechanical) description, where the size of a particle extends into areas that are not causally connected (i.e., extends beyond the event horizon), leads to a loss of causality. Therefore, it makes sense to initially leave aside concepts like time and space for the description or propagation of a state, until causality is re-established after a longer time and over larger areas.

---

[26] The term de-Broglie wavelength was coined by Louis de Broglie in 1924: Any form of matter or energy has wave character. This property is a cornerstone of quantum mechanics and is known as particle-wave duality. Experimentally, this oddity was first demonstrated for electron beams, whose diffraction behavior exactly resembles that of light beams. For macroscopic objects (i.e., those of everyday life), the wavelengths are much too small to have any detectable effects.

To derive the relevant Planck parameters:

(1) $2\pi\times$horizon-radius: $2\pi R_H = 2\pi \cdot Gm/c^2$
(2) de-Broglie wavelength: $\lambda_{\mathrm{dB}} = 2\pi\hbar c/(mc^2) \doteq 2\pi R_H$
(3) (1) = (2) $\Rightarrow$ Planck mass: $m_{\mathrm{Pl}} = \sqrt{\hbar c/G}$
(4) Planck length (from (2)): $\lambda_{\mathrm{Pl}} = \hbar c/(m_{\mathrm{Pl}}\, c^2)$
(5) Planck time (from (4)): $t_{\mathrm{Pl}} = \lambda_{\mathrm{Pl}}/c$
(6) Planck temperature (from (3)): $T_{\mathrm{Pl}} = m_{\mathrm{Pl}}\, c^2/k_B$

Note: In equation (1) we have quoted the horizon radius of a maximally rotating Black Hole, which is one half the corresponding radius of a non-rotating Black Hole of the same mass.

For easy calculation in SI units:
$$c \simeq 3 \cdot 10^8, \quad \hbar c \simeq \pi \cdot 10^{-26}, \quad G \simeq 6.6 \cdot 10^{-11}, \quad k_B \simeq 1.4 \cdot 10^{-23}$$

The numerical values for the Planck sizes are then:
$m_{\mathrm{Pl}} \simeq 1.2 \cdot 10^{19}$ GeV/c$^2 \simeq 22\mu$g, $\lambda_{\mathrm{Pl}} \simeq 1.6 \cdot 10^{-35}$ m,
$t_{\mathrm{Pl}} \simeq 5.4 \cdot 10^{-44}$ s, $T_{\mathrm{Pl}} \simeq 1.4 \cdot 10^{32}$ K.

Since in the earliest Universe, i.e., even up to the inflationary phase, massive particles do not yet exist, most of the current theories are based on thermal approaches that can easily be derived from the Einstein equations or likewise from the Friedmann equations of motion.

The initial state may certainly have many unknown parameters, nonetheless it is characterized by an extremely large energy density of space-time. At this time, the space does not contain masses yet, so it is equivalent to a de-Sitter universe with a cosmological constant (see page 65). The state propagates as a function of time towards a phase transition; however, this phase transition is delayed, quite similar to a supercooled liquid[27]. The

---

[27] A supercooled liquid is one that is (carefully) cooled down below its actual freezing or melting point. In the case of water, this is possible down to about $-10°$ C, before it crystallizes into ice and suddenly expands. Gallium (melting point $+30°$ C) can be supercooled to about $-40°$ C, before it crystallizes. During supercooling the density does not change. [Cf. D. Frekers, et al. *A technique for the preparation of thin self-supporting metallic gallium targets*, Nuclear Instruments and Methods, Physics Research A 621, 704 (2010)].

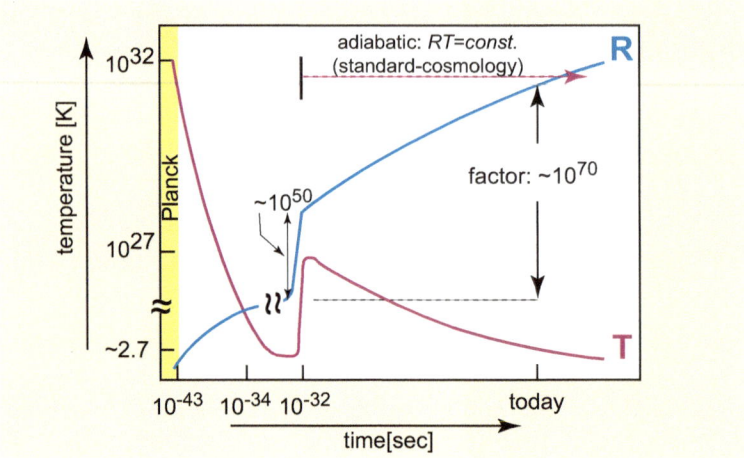

Fig. 6.16: The cosmic evolution with the initial inflation. In the pre-inflationary phase, the space-time first expands quasi-adiabatically and finally reaches a false vacuum because of the delayed phase transition. In this state, the density does not change, which then leads to an exponential expansion according to Friedmann's equations, where the Universe expands by about $40 - 70$ orders of magnitude depending on the length of the delay. In the subsequent phase transition, elementary particles crystallize out, and the latent heat leads to a drastic rebound in temperature. Then, in the following quiescent post-inflationary phase, the entropy of the system remains constant and the expansion becomes fully adiabatic (i.e., the Universe stays as a closed system with no gain or loss of heat from or to the outside).

theory of inflation refers to this as a transition into a "false vacuum". Since the energy density does not change during the transition into this false vacuum, this is equivalent to a negative pressure (or tension) in Friedmann's equations and, according to these equations, leads to an exponential expansion by $40 - 70$ orders of magnitude (inflation), where these numbers are approximately derived from present-day observables. "Negative pressure" signifies that the expansion is not compensated by a counter-force against which the pressure would have to do work, i.e., the space expands unhindered. As part of the phase transition, the elementary particles "crystallize" out. The resulting latent heat [28] is transferred to the nascent particles, leading to a temperature rise almost back to the Planck temperature.

---

[28] Latent heat is generated during a phase transition from a low to a higher order state. For example, during the crystallization of water to ice, $333.5\,\mathrm{kJ/kg}$ ($=79.7\,\mathrm{kcal/kg}$) of heat is released that must be dissipated to produce ice. In the case of gallium, this heat is $80\,\mathrm{kJ/kg}$ ($=19.1\,\mathrm{kcal/kg}$).

From now on, the state variable referred to as "entropy", which in thermo-dynamics describes the interplay between energy and temperature, plays a decisive role. A phase transition is characterized by a change of entropy. In a transition from a low-ordered (example: liquid) to a higher-ordered state (example: crystallized solid), the entropy decreases and the amount of heat given off increases, and vice versa. The Planck energy (or Planck temperature) is the initial anchor point for a maximum possible temperature. After the end of the inflationary phase, though, the entropy S of the Universe remains constant until the present time. It is given by the product of the scale parameter R and the temperature T ($S \propto RT = $ const.), i.e., with increasing expansion R the temperature T decreases. This is in accordance with the observations. The situation is roughly visualized in Fig. 6.16.

The scenario played out here solves in an elegant way the fine-tuning problem presented at the beginning, without inventing fundamentally new physics. The problematic back calculation to time $t = 0$ is omitted, and the critical density $\Omega_{crit} \simeq 1$ occurs automatically with the degree of flatness. However, it pushes the fine-tuning problem into the future. Our current Universe emerged in this model from an even smaller area than the initial event horizon. It is therefore conceivable that areas, which currently are still far outside our event horizon, will enter into it at some time in the future, where flatness and homogeneity are by no means guaranteed. Estimates assume an approximate time scale of 10^{20}–10^{21} seconds, which is roughly a thousand times the age of the present Universe. If the density is then larger than the critical density, the Universe comes to a rapid collapse — that would then be the proverbial "End of the World", and the most interesting epochs, which are outlined on pages 50 to 52 and in Fig. 5.1, would not take place — unfortunately. But maybe there will be another chance next time someone tries.

Chapter 7

Nucleosynthesis in the Early Universe

Also at the beginning of this chapter, very simple questions, those which result from experimental observations of astronomy and astrophysics, shall again be asked first:

▶ Where do the chemical elements come from? — could it be that they were formed during the "Big Bang" phase?

▶ The Universe consists of about 75% hydrogen (isotope ^{1}H) and 25% helium (isotope ^{4}He), the rest is totally negligible with an abundance less than $< 1‰$ — how to understand these numbers?

▶ Why is there almost no deuterium (isotope ^{2}H or D), although stable, and why is there almost no helium-3 (^{3}He), although also stable? — the fraction of deuterium in the Universe is only $1/40,000$ that of the Hydrogen ^{1}H and the fraction of ^{3}He is only $1/20,000$ that of helium-4 (^{4}He) — again, how to understand these numbers?

▶ Where do the neutrons come from, which are indispensable for the synthesis of heavy elements, given that not even for the deuterium or the helium-3 an effective synthesis seems possible.

▶ Finally, why almost all available neutrons seem to be consumed for the synthesis of ^{4}He, which after all makes up 25% of the elemental abundance?

© The Author(s), under exclusive license to Springer-Verlag GmbH, DE, part of Springer Nature 2025
D. Frekers, P. Biermann, *Universe, Neutrinos, Stars and Life*, https://doi.org/10.1007/978-3-662-70729-6_7

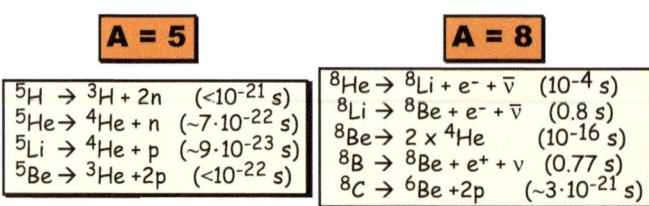

Fig. 7.1: The situation of nuclei with mass number A=5 and A=8. For A=5 there are no bound states. For A=8 the decay times are all smaller than 1 second.

Until the late 1940s, the common narrative said that all naturally occurring elements were synthesized "somehow" during the "Big Bang" phase. This notion was, of course, due to ignorance of the microwave background radiation, which had not been discovered yet, but also due to ignorance of the significance of nuclear reactions that play a role in this. And furthermore, the physics of stellar explosions, like supernovae, was completely in the dark at that time. With today's knowledge about the background radiation and given the immense body of experimental data on nuclear reactions of any kind, the above statement that the "Big Bang" is responsible for the synthesis of all elements is completely untenable.

Why? — Answer: There are four key features in nuclear physics, established by Nature, which prevent a significant nucleosynthesis beyond $A = 4$ (i.e., beyond ^{4}He) from occurring during the "Big Bang" phase. These are:

1. the extremely small binding energy of only 1.11 MeV per nucleon (2.22 MeV in total) of the deuteron — thus, the deuteron is among all stable element-isotopes the one with the lowest binding energy,
2. the high binding energy of ^{4}He of 7.07 MeV per nucleon (28.3 MeV in total), which is the highest in this lower mass region,
3. the fact that there are no stable nuclei with mass number $A = 5$ or $A = 8$ (see Fig. 7.1), and
4. the half-life of the free neutron of 609.6 ± 0.6 seconds.

Minor changes to any of these parameters would have dramatic effects on the further evolution of the Universe up to an early galaxy collapse or even the complete absence of structure formation, i.e., no galaxies, no stars, no Earth. It is therefore appropriate to take a closer look at the aspects that have such direct implications to our existence.

7.1 Neutrons come and go

Neutrons play a decisive role in the phase of early nucleosynthesis — without them, nucleosynthesis is simply not possible. But then, where do they come from, and how many of them are there compared to the number of protons?

We click into the Early Universe at a time when the temperature is still much higher than the equivalent mass difference between proton and neutron, the latter of which is 1.293 MeV, i. e.:

$$k_B T \gg \Delta M_{np} = M_n - M_p = 1.293 \text{ MeV}$$

One may recall that at this time photons, electrons, positrons, and neutrinos are the main constituents of the Universe, which are by number about 10^{10} times more abundant than, for instance, protons. All components are in thermal equilibrium due to their interaction with each other.

In this environment, there are now four decisive reactions that convert proton and neutron into each other. In addition, at high enough temperatures $(k_B T \gg 10 \text{ MeV})$ the reaction rates of the two conversion directions are nearly equal. These four reactions are mediated by the «*weak*» interaction and can be easily calculated with high accuracy at any time of the cooling down of the Universe:

$$
\begin{array}{lll}
(1) & p + e^- \longrightarrow n + \nu_e & \text{E} = -0.782 \text{ MeV} \\
(2) & n + e^+ \longrightarrow p + \overline{\nu}_e & \text{E} = +1.293 \text{ MeV} \\
(3) & n + \nu_e \longrightarrow p + e^- & \text{E} = +0.782 \text{ MeV} \\
(4) & p + \overline{\nu}_e \longrightarrow n + e^+ & \text{E} = -1.293 \text{ MeV}
\end{array}
$$

To note: Reactions (3) and (4) are the reverse reactions of (1) and (2); futher, the specified energies in (1) and (4) must be supplied by the temperature bath, whereas the energies in (2) and (3) are released to the temperature bath.

Protons and neutrons are thus in thermal equilibrium. However, "thermal equilibrium" does not mean that the numerical neutron/proton ratio (N_n/N_p) settles exactly at the value 1 (i.e., at 50/50). This ratio is given in classical thermodynamics by the Boltzmann distribution, which accounts

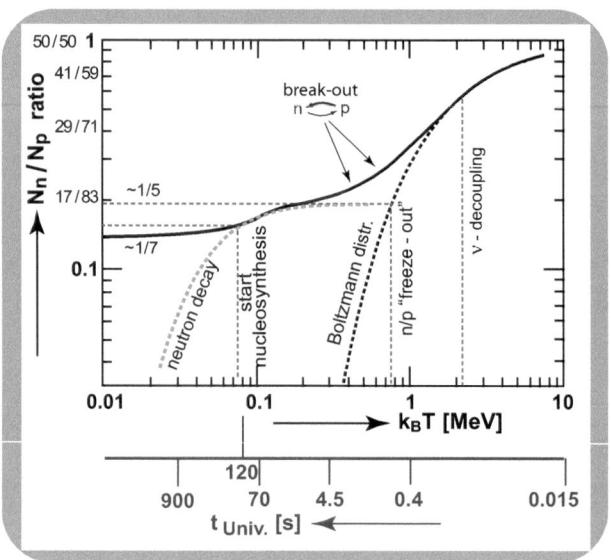

Fig. 7.2:   Calculated neutron/proton ratio N_n/N_p in the Early Universe as a function of temperature and age. Since the sum of protons and neutrons remains constant, the relative percentages are indicated at the ordinate as a guide. Also shown are the individual stages leading to the neutron/proton ratio, which are: the neutrino decoupling at about 0.26 seconds, the purely thermodyamic "freeze-out" process for neutrons at about 780 keV, the neutron decay, and finally the start of the fusion $n + p \rightarrow d$ at about 75 keV and the subsequent stabilization of the neutron/proton ratio at about 0.139 for all times.

for the fact that the masses of the two nucleons M_n and M_p are different:

$$\frac{N_n}{N_p} = \left(\frac{M_n}{M_p}\right)^{3/2} e^{-\frac{\Delta M_{np}}{k_B T}}$$

Therefore, if the temperature equivalent energy $k_B T$ comes close to the mass difference ΔM_{np}, the neutron/proton ratio N_n/N_p drops exponentially. This still assumes, though, that the interaction strengths, which convert neutron and proton into each other, are the same for both conversion directions. This is now getting to be less and less the case.

At $k_B T \simeq 2\,\text{MeV}$ (i.e., after about 0.26 seconds) the electron neutrinos decouple, and the above reactions (3) and (4) will cease. And finally at $k_B T \leq 1\,\text{MeV}$, the different energetics of the remaining reactions (1) and (2) and the associated different interaction rates start to gain importance. They cause the neutron/proton ratio N_n/N_p to break out of equilibrium

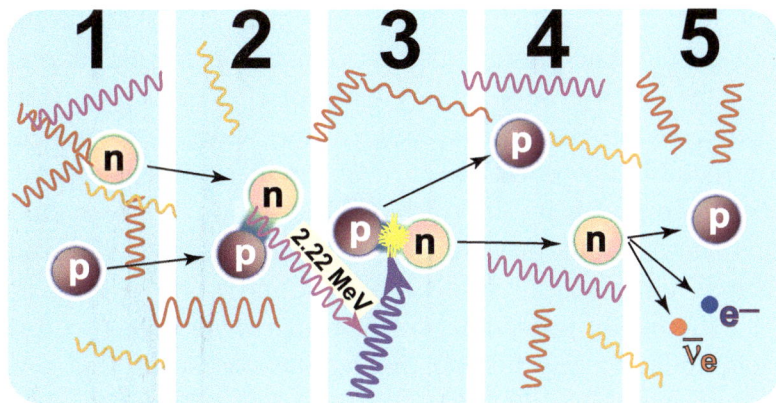

Fig. 7.3:   Depicting the *futile!* attempt of a proton and a neutron to fuse to form the deuteron. Proton and neutron find themselves in a photon bath (1) with about 10^{10} times more photons. Fusion to a deuteron (2) releases a photon of about 2.22 MeV, but almost instantaneously another photon from the thermal bath is absorbed (3), and the deuteron dissociates back into a proton and neutron (4). The neutron eventually gives up and decays with its innate half-life into a proton, an electron, and an anti-neutrino (5). ***Result: no deuteron!***

and eventually to freeze at a ratio $17/83$ ($\simeq 1/5$) after a few seconds. A decisive factor for this relatively low value is also the fact that electrons and positrons, which kept the reactions (1) and (2) going after the neutrino decoupling, only annihilate — and thereby disappear — until much later, i.e., after about 26 seconds via $e^- + e^+ \to 2\gamma$. In Fig. 7.2 these various phases are graphically illustrated.

A minor side effect may be worth mentioning: The extra photons created by this $e^- + e^+$ annihilation process cause an increase of the temperature of the Universe by about 40%. Hence, the photon temperature of the Universe will from now on lie above the neutrino temperature for all times to come, e.g., by today $T_{\text{photon}} = 2.725\,\text{K}$ (as measured) and $T_{\text{neutrino}} = 1.94\,\text{K}$ (as predicted).

With the freeze-out of the neutrons, a cosmic race against time begins. In principle, the first step for the formation of elements, namely the fusion of proton and neutron to a deuteron, could start now. But even though the temperature has meanwhile dropped far below the deuteron's binding energy of 2.22 MeV, initially any such attempt is doomed to fail (see Fig. 7.3). The reason is the enormous, almost 10^{10}-fold excess of photons from the background radiation, which leads to the fact that the intensity in the high-energy part of the energy distribution is for a long time sufficient to instantaneously destroy (dissociate) each of the deuterons formed. Only after the Universe has cooled to about 75 keV, which is around 140 seconds, does an appreciable fusion rate occur. In the meantime, the neutron decays back to the proton with its innate half-life of about 610 seconds, and after 140 seconds the neutron/proton ratio has dropped to 0.15. However, once the neutron is captured in a deuteron bound state, decay is inhibited. Fortunately, the fusion process is so fast that the neutron/proton ratio does not drop significantly any further during the course of this process. The ratio settles at around 0.139 ($\simeq 1/7$) after a few minutes.

The fusion of a neutron and a proton to a deuteron is only a short interlude. Because, once the deuteron production has started, the next nucleosynthesis process immediately sets in (see further below). This step consists of various reaction sequences, which are interconnected through a complex network of nuclear reactions. In the end almost all neutrons end up bound in ^{4}He (i.e., 2 neutrons to 2 protons each). And given the initial neutron/proton ratio of 0.139, the value for the ^{4}He mass ratio in the Universe that comes out of this network is 24.4%. This is to be compared with the experimentally measured value of $24.49 \pm 0.04\%$ [1] — a truly outstanding agreement and a triumph of the "Big Bang" model.

Here in mathematical terms to evaluate the ^{4}He mass ratio from the neutron to proton ratio of 0.139 assuming all neutrons get bound into ^{4}He:

For a neutron/proton ratio of

$$N_n/N_p = 0.139$$

the number ratio of ^{4}He/^{1}H (i.e., helium/hydrogen) is

$$X_4 = \frac{N(^4\text{He})}{N(^1\text{H})} = \frac{N_n/2}{N_p - N_n} = \frac{1}{2}\left[\frac{1}{N_p/N_n - 1}\right] = 0.081\,.$$

[1]   E. Aver, et al., *The effects of He I λ10830 on helium abundance determinations*, Journal of Cosmology and Astroparticle Physics 07, 11 (2015), see also arXiv:1503.08146v1 [astro-ph.CO].

The numerator results from the fact that two neutrons are bound in each ^{4}He, and the difference in the denominator that the number of protons in the ^{4}He is equal to the number of neutrons. The result says that a total of about 8.1% of all baryonic particles in the Universe are ^{4}He particles.

From this the mass ratio is calculated

$$Y_4 = \frac{M(^4\text{He})}{M(^4\text{He}) + M(\text{H})} = \frac{4X_4}{4X_4 + 1} = 0.244 \,,$$

by using that the mass of ^{4}He is approximately 4 times the mass of the neutron — the difference is only 0.7%.

7.2   Beginning and end of nucleosynthesis

The low binding energy of the deuteron was the reason that the Universe had to cool down to 75 keV before a fusion to the deuteron could take place, and by making deuterons as well as helium, the reservoir of free neutrons was rapidly being emptied out. Any further fusion processes to build heavier elements beyond $A = 4$ would now require progressively more charged particle reactions. However, the temperature for these has already dropped far too low. To note, the Coulomb (electrostatic) repulsion potentials, which these particles must overcome to be able to fuse — i.e., about 1 MeV for $d + d$, about 2 MeV for d+He, and about 3 MeV for He+He — are way too high compared to their kinetic energies at this temperature. Classically, the making of the Universe has come to an end. For a little while, though, the quantum mechanical tunnel-effect comes to the rescue, but also this goes rapidly and exponentially to zero.

Finally, after about 10 minutes, the Universe has cooled by another 40 keV, the particle density has also dropped below a critical value, and all nuclear reactions come to a halt. The short-lived unstable elements decay, as there are ^{3}H (tritium): $t \longrightarrow ^3$He (12.3 years) or ^{7}Be: ^{7}Be$\longrightarrow ^7$Li (53 days). Long-lived ones do not exist. By their decay, they still change the elemental composition for a short time before the latter is finally frozen and exposed to an observation 13.81 billion years later.

The most important fusion reactions are summarized here:

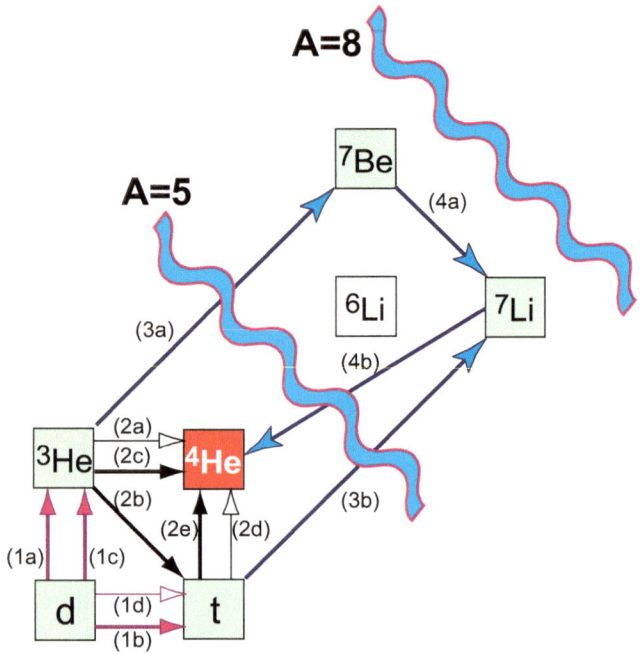

primary reactions

(1a) $d + d \longrightarrow {}^3\mathrm{He} + n$

(1b) $d + d \longrightarrow t + p$

(1c) $d + p \longrightarrow {}^3\mathrm{He} + \gamma$

(1d) $d + n \longrightarrow t + \gamma$

secondary Reactions

(2a) ${}^3\mathrm{He} + n \longrightarrow {}^4\mathrm{He} + \gamma$

(2b) ${}^3\mathrm{He} + n \longrightarrow t + p$

(2c) ${}^3\mathrm{He} + d \longrightarrow {}^4\mathrm{He} + p$

(2d) $t + p \longrightarrow {}^4\mathrm{He} + \gamma$

(2e) $t + d \longrightarrow {}^4\mathrm{He} + n$

tertiary reactions

(3a) ${}^3\mathrm{He} + {}^4\mathrm{He} \longrightarrow {}^7\mathrm{Be} + \gamma$

(3b) $t + {}^4\mathrm{He} \longrightarrow {}^7\mathrm{Li} + \gamma$

back-reactions

(4a) ${}^7\mathrm{Be} + n \longrightarrow {}^7\mathrm{Li} + p$

(4b) ${}^7\mathrm{Li} + p \longrightarrow 2 \times {}^4\mathrm{He}$

Fig. 7.4:   Network of reactions in the Early Universe. Of the nearly 400 reactions that have a possible influence on early nucleosynthesis, only the strongest are shown here. To note: (i) no stable masses at A=5 and A=8; (ii) there are no outgoing arrows from the ${}^4\mathrm{He}$; (iii) with the production of ${}^7\mathrm{Be}$ via the channel (3a), its destruction follows via ${}^7\mathrm{Li}$ (4a) back to the ${}^4\mathrm{He}$ (4b); (iv) the same is true for the production of ${}^7\mathrm{Li}$ via the channel (3b) and its return to the ${}^4\mathrm{He}$, (v) there is no significant generation of ${}^6\mathrm{Li}$. Note also the color coding of the reaction types. Open arrows represent comparatively weak reactions.

Surprisingly, all reactions lead to ${}^4\mathrm{He}$, and indeed, at the end of the nucleosynthesis, ${}^4\mathrm{He}$ is by far the most abundant element after hydrogen. This is due to the extraordinarily high binding energy of ${}^4\mathrm{He}$ that makes it impossible to subsequently break up ${}^4\mathrm{He}$ again. The modest attempt to bridge the mass A=5 in the network by means of the reactions (4a) and (3b) shown in Fig. 7.4 leads for the most part back to ${}^4\mathrm{He}$. A breakout from the network by a jump over the mass A=8 using ${}^7\mathrm{Li}$ or ${}^7\mathrm{Be}$ as starting points (not shown here) also remains utterly unsuccessful. The resulting abundances $^9\text{Be}/^1\text{H} \approx 10^{-18}$ and $^{10}\text{B}/^1\text{H} \approx 10^{-21}$ are far too low to initiate any further fusion reactions. This makes it unequivocally clear that the heavy elements beyond A=7 must be formed in a completely different astrophysical environment.

At present, the theoretical network reactions are extended to several hundred single reactions, and even small and smallest reaction rates are being considered. Thus, by playing with parameters, one can pursue the questions: (1) how do the element abundances change as a function of the baryon-to-photon ratio — a parameter, which is, of course, experimentally determined by the background radiation — and (2) are the element abundances calculated in this way compatible with experimental observations? Remember, the baryon-to-photon ratio was the decisive factor for the significantly delayed formation of the deuteron. Fig. 7.5 shows the results of such calculations. For this purpose, the actually observed element abundances as well as the crossing points with the baryon-to-photon ratio of 6×10^{-10} determined by the Planck probe are plotted. The agreement for the elements D, ^3He, and ^4He is simply impressive and represents a highlight of the

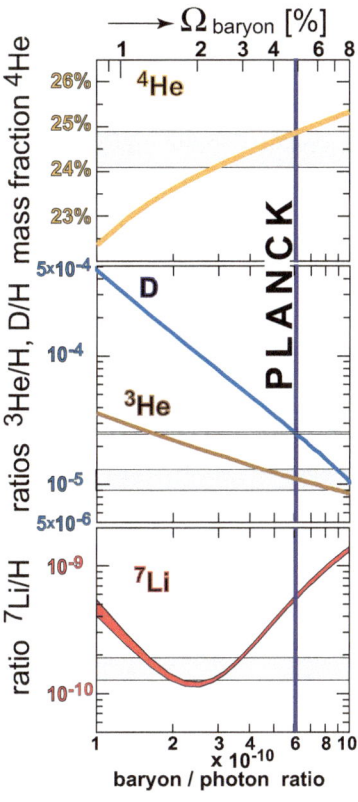

Fig. 7.5: Nucleosynthesis in the Early Universe. Shown are theoretical network calculations as a function of the baryon-to-photon ratio and the baryon density (scale at top in units of the total density of the Universe). Experimental reaction rates are included in these calculations, and their errors are indicated by the line thicknesses. The actual measured element abundances are given by the horizontal lines together with their error widths. The baryon-to-photon ratio and the baryon density, both measured by the Planck probe are connected as vertical lines. The agreement of theory and experiment is remarkable – exception: ^7Li (see text).

"Big Bang" model. However, the discrepancy for ^{7}Li, whose experimentally determined abundance is a factor of 3.5 lower than that predicted by the "Big Bang" model, is all the more surprising. The situation becomes even more confusing for the stable isotope ^{6}Li (not shown here), whose element abundance ^{6}Li/H in the calculations is about 1.3×10^{-14}, whereas observational values come out to be larger by a factor of 500.

A comment on Li abundances: How to determine primordial elemental abundances in observational astronomy is outside the scope of this book. Nonetheless, a brief comment may be useful. In most cases, atomic absorption or emission lines from intergalactic clouds are studied spectroscopically. These objects should preferentially be localized at a high redshift ($z \geq 2$, i.e., at a distance of ≥ 10 billion light-years) to ensure that the elemental abundances measured there are a correct reflection of the "Big Bang" and no significant mixing or contamination with stellar material has taken place yet. For the explanation of the lithium discrepancy, this is an important consideration because the initial lithium abundances are extremely low, and even minute amounts of extra lithium injected into these clouds could lead to misinterpretation. Indeed, recent cosmological models suggest that the lithium abundances may have already been altered in the earliest stages of stellar evolution and that they may not, as assumed, be a reflection of the "Big Bang". Thus, the above quoted lithium discrepancy could well be unfounded, and any of the alternative explanations, which either require a change of various "adjusting screws" in primordial synthesis — which in almost all cases leads to conflicts with experimental data — or bring into play new, unknown physics[2], could simply be skipped. Interesting in this

---

[2] A good summary of this topic can be found in
Alain Coc, Elisabeth Vangioni, *Primordial Nucleosynthesis*, International Journal of Modern Physics E 26, 1741002 (2017), see also same authors, same title, arXiv:1707.01004 [astro-ph.CO],
— and Alain Coc, et al., *New reaction rates for improved primordial D/H calculation and the cosmic evolution of deuterium* Physical Review D 92, 123526 (2015), see also same authors, same title, arXiv:1511.03843 [astro-ph.CO],
— and Brian D. Fields, *The Primordial Lithium Problem*, Annual Reviews of Nuclear and Particle Science 61, 47 (2011), see also same authors, same title, arXiv:1203.3551v1 [astro-ph.CO],
— and K. Lind, et al., *The lithium isotopic ratio in very metal-poor stars*, Astronomy & Astrophysics 554, A96 (2013).

context are, for instance, astronomical observations described in reference [3]. The lithium abundances measured in the interstellar and presumed primordial nebula of the Small Magellanic Cloud, which is, after all, at our doorstep, agree with those from the "Big Bang" model. Nonetheless, a definitive and confirmed explanation for the lithium discrepancy is still pending.

What would be, if...?

It is certainly not the task of physics to speculate about things that do not exist, or even to describe events that never occurred. Nevertheless, it is amusing and impressive at times to ask the question "what would be, if...?" especially when trying to get an idea of how precisely and accurately things around us are possibly fine-tuned. This is especially true for the initial parameters that played a role in the creation of the Universe. But beware: such elaborations remain fundamentally speculative because no one can ultimately account for the effects in their entirety, even if only a few of these "adjusting screws" are turned. The question of whether the Universe is indeed an extremely unstable entity, or whether it is by and large self-stabilizing through the complex and convoluted interplay of all "adjusting screws" can hardly be answered and leads us directly back to the "anthropic principle" already cited on page 58.

Nevertheless, we will ask these nonphysical and nonsensical questions:

What would be,

1. if the neutron mass were 0.02% (two 10,000ths) larger?

(a) the half-life of the neutron would be 111 seconds instead of the current 610 seconds,

(b) deuterons could not be created in significant numbers because about 90% of all neutrons would have already decayed at the beginning of the nucleosynthesis,

(c) the binding energy of the deuteron would be even lower, which would push a possible nucleosynthesis even further back,

---

[3] J. C. Howk, et al., *Observation of interstellar lithium in the low-metallicity Small Magellanic Cloud*, Nature 489, 121 (2012).

(d) since there wouldn't be any deuterons, there would also not be a significant helium fraction in the Early Universe,

(e) there would be about 25% more stars in the sky (because of 25% more hydrogen available),

(f) since almost all lithium on Earth is primordial, there would be virtually no lithium on Earth — and thus no Li batteries.

2. if the neutron mass were 0.05% (five 10,000ths) smaller?

(a) the half-life of the neutron would now be about 10,000 seconds instead of 610 seconds,

(b) the decay of neutrons would hardly play a role — all neutrons (now about 20 – 30%) would fuse to deuterons,

(c) the binding energy of the deuteron would be larger, which would shift the nucleosynthesis to shorter times and make it even more effective,

(d) there would be significantly more helium formed — the mass fraction of currently 25% would increase to about 50%,

(e) there would be about 50% fewer stars, which would be in a Sun-like hydrogen burning cycle with long lifetimes,

(f) yet at the beginning (i.e., in the first 1 – 2 billion years) there would be many more bright stars, as these would be in the He-burning phase; and because their lifetimes would be much shorter, they would have already perished by turning into White Dwarfs,

(g) the formation of a Sun with planet Earth would be less probable by at least a factor 2 — every second sun would not exist.

3. if the baryon/photon ratio given by matter-antimatter annihilation in the earliest Universe were not 6×10^{-10} but rather 6×10^{-9}, i.e., an order of magnitude larger?

(a) the hydrogen and helium-4 proportions would remain about the same, the element synthesis would merely shift in favor of the heavier elements, i.e., lithium-7 would now be about 100 – 200 times more abundant on Earth, deuterium about 4 to 5 orders of magnitude rarer,

(b) there would be about 10 times more hydrogen in the Universe and thus about 10 times more stars in the sky — it is not clear if this would be enough for an early galaxy collapse, though it is quite conceivable. An Earth wouldn't have had a chance.

4. if the baryon/photon ratio from the same process as above, were not 6×10^{-10} but rather 6×10^{-11}, i.e., an order of magnitude smaller?

(a) in this scenario, the fusion of elements is slowed down, there would be about $100 - 200$ times more deuterium, a factor of 10 more helium-3, and the mass fraction of helium-4 would be about 15% instead of the 25% today,

(b) overall, there would be almost a factor of 10 fewer baryons in the Universe because baryons and anti-baryons would have been created more symmetrically in the early formation phase; the later annihilation process would have been much more effective,

(c) whether under these conditions any significant structure formation with subsequent stellar and galaxy evolution, and thus a planet Earth, could have occurred at all, is at least doubtful.

5. if the «*weak*» interaction were by about 10% "less weak"?

(a) in the Early Universe, the freeze-out of the neutron/proton ratio would be further delayed and the number of free neutrons needed for initial fusion to deuterons would be reduced by about 50%,

(b) the half-life of the free neutron would now be about 550 s instead of 610 s, with the added effect that even fewer neutrons would be available for deuteron fusion,

(c) the mass fraction of helium-4 would now be less than 10%,

(d) the number of stars in the sky would be about 20% higher (because about 20% more hydrogen present),

(e) the lifetime of stars (and therefore also that of the Sun) would be about $10 - 20\%$ shorter,

(f) supernova explosions, in which processes of the «*weak*» interaction play an important role, would be much more frequent and much more violent so that the nucleosynthesis of the heavy elements (i.e., all elements with mass number A $>$ 4) would take a completely different path,

(g) the radiation power of the stars (and therefore also that of the Sun) would be about $10-20\%$ higher,

(h) the Earth would be located outside the habitable zone of the Sun; Earth would need another solar system.

One can now ponder all the conceivable combinations of these effects. However, it becomes clear quite quickly that one plunges into an evidenceless nirvana. Therefore, we want to quickly return to "what is" and not further speculate about "what would be if".

Chapter 8

The Microwave Background Radiation

8.1   The 380,000 year mark

A Universe, which presently expands and becomes larger, must — as already mentioned at the beginning of this book — have been smaller in the past. Since the baryonic, i.e., the "normal" interacting part of this Universe consists almost to 100% of a gas, namely hydrogen and helium, this gas must have been more compressed in earlier times and consequently been at a higher temperature. This reasoning can be followed further back in

© The Author(s), under exclusive license to Springer-Verlag GmbH, DE, part of Springer Nature 2025
D. Frekers, P. Biermann, *Universe, Neutrinos, Stars and Life*, https://doi.org/10.1007/978-3-662-70729-6_8

time up to the era that marks the end of inflation, i.e., even long before helium came into existence. In fact, as long as no heat exchange has taken place, it is largely irrelevant whether in the course of evolution components have transformed, decayed or decoupled from the thermal bath, such as the neutrinos in the first few 100 ms. Such transformation effects are captured in thermodynamics by the state variable entropy, which will not be discussed here now.

All things in our environment are subject to « *electromagnetic* » interaction. And with temperature, or synonymously heat, comes the so-called electromagnetic temperature radiation (i.e., photon emission). The properties of this radiation are described by Planck's radiation formula for "blackbody radiators", where the term "black" means that there should neither be characteristic self-emission nor should there be any dynamics of the constituents deviating from thermodynamics. The radiator radiates and cools in this process with the property: the higher the temperature, the shorter the wavelength, i.e., the higher the energy of the radiation the more effective the cooling. The radiation then has a typical bell-shaped intensity curve, which is depicted in Fig. 8.1 and also in Fig. 8.4, whereby the latter represents an experimental result. It sounds contradictory, but our central star, the Sun, is an example of an almost ideal "blackbody radiator".

With the knowledge that the Universe expands, these very elementary facts were known, though, in the literature, they were not discussed in connection with the Universe. This was probably so because the general opinion at that time was that such a radiation, after approx. 13 billion years, would no longer be measurable or that no useful information could be extracted from it anyway. This should turn out to be a **big mistake** in the time that followed.

In fact, the temperature radiation is witness to a series of epochs in the Early Universe and each epoch is able to put its own imprint into the radiation, although the imprint may fade over time as a result of thermalization.

As the Universe expands, the temperature of the radiation decreases. However, the radiation remains in contact with the constituents of the Universe for a long time, even well beyond the phase of early nucleosynthesis. Therefore, if the nucleons (baryons) in this early phase still have knowledge about

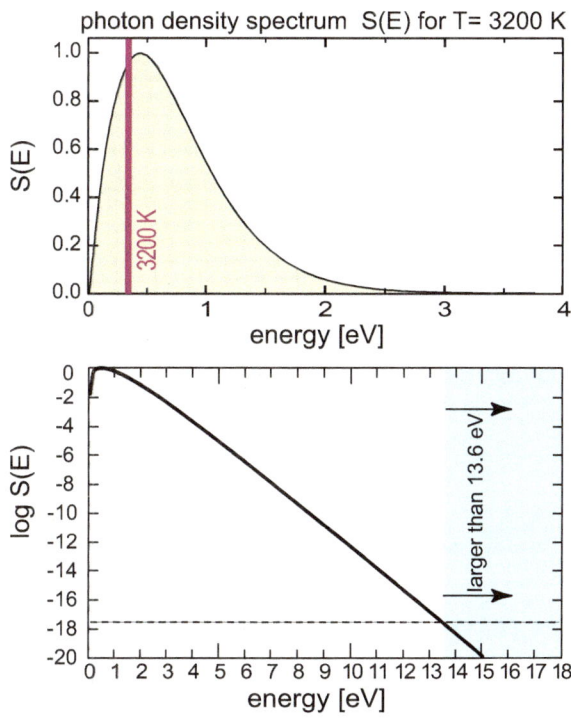

Fig. 8.1:   Planck's "blackbody" radiation spectrum for a temperature of 3200 K. The spectrum S(E) is arbitrarily normalized to a maximum value of 1.   (For orientation: the temperature of 3200 K corresponds to an energy of $k_B \cdot T \simeq 0.28\,\mathrm{eV}$.)

The lower part shows the spectrum up to the radiation energy of 18 eV in logarithmic form.   Only photons above an energy of 13.6 eV are capable of ionizing hydrogen.   Their relative fraction is below 10^{-16} at this temperature.

the early and violent phase of the Universe, then there ought to be no question that this information must have been transferred like a fingerprint to the profile of the temperature radiation.

After approx. 380,000 years, there is a break!

Because — after about 380,000 years the Universe has cooled down to about 3200 K. The fraction of photons in the high-energy part of the Planck spectrum, i.e., the fraction that lies above the binding energy of the electron in the hydrogen atom of 13.6 eV, is from now on too small for continuously ionizing hydrogen (see Fig. 8.1). The proton and electron finally recombine to form the hydrogen atom, i.e.,

$$p + e^- \quad \longrightarrow \quad \mathrm{H} \ + \ 13.6 \text{ eV} \qquad \text{recombination}$$
$$\mathrm{H} + \mathrm{h}\nu \quad \not\longrightarrow \quad p + e^- \qquad \text{re-ionization no longer possible}$$

This epoch lasts about 30,000 years and ends the era of the interaction of thermal photons with the Universe. The photons decouple, they move from now on in straight lines and are subject only to the cosmological redshift. The Universe has become transparent. The epoch is called "recombination era", also paraphrased with the term "last scattering surface". This last scattering surface surrounds us in the form of an opaque spherical surface, which can be seen from inside, and which is 13.8 billion light-years away and about 20,000 light-years thick [1].

All properties deviating from thermodynamics, which had been encoded into the temperature radiation, are henceforth frozen like a fingerprint, and moreover, the correlation between the size of the fingerprint and the baryonic density fluctuations in the Universe, which after all caused this imprint, remains unchangeably engraved.

8.2   The discovery of the radiation

Interestingly, the cosmic background radiation was discovered completely by accident in 1964 by Arno Penzias and Robert Wilson. Both were employed by the research laboratory of the Bell telephone company in Holmdel, New Jersey. They had been given the task of developing a new type of a high-sensitivity receiving antenna in the microwave range for the NASA ECHO satellite project. The ECHO-2 satellite (a successor to the less successful ECHO-1 and ECHO-1A) was a balloon satellite 41 meters in diameter that was launched into a 1200 km high Earth orbit on January 25, 1964, where it deployed to its full size [2]. As a passive satellite, it was intended to reflect microwaves from an Earth-bound transmitting station back to Earth, and thereby to enable long-distance communications and at the same time to cover large areas on Earth. The satellite transmitted at a wavelength of 7.35 cm, which is just within the window of maximum transmittance of the Earth's atmosphere (see Fig. 8.3).

---

[1] The "last scattering surface" has an analogy to the solar surface. For photons generated inside the Sun, it takes millions of years to reach the surface due to continuous scattering. Once there, they travel in a straight line. The scattering disallows a view into the Sun's interior.

[2] The older generation may remember that the ECHO satellites were visible to the naked eye from Earth in darkness and clear skies.

Fig. 8.2: This photo shows Arno Penzias (foreground) and Robert Wilson in front of their antenna with its 6 m long horn-shaped aperture.

Credit: National Park Service, copyright notice: "No claim to original U.S. Government works."

During commissioning of the antenna (at liquid-helium temperatures of $-269\,°C$), Penzias and Wilson encountered a unexpected noise, which was independent of direction and whose amplitude was about 100 times bigger than what they had anticipated as an ambient background noise. After a painstaking and futile search for the cause and after eliminating all known interference effects such as radar and radio, they were finally at a loss but nevertheless certain that an inherent noise of the antenna could definitely be ruled out.

Sixty kilometers away, at Princeton University, there was the group of Robert Dicke, Jim Peebles, and David Wilkinson, who were planning a dedicated experiment for the detection of the predicted cosmic microwave background radiation. Penzias and Wilson learned of these activities by chance and only then realized what a discovery they had made. Together they decided to have the experimental data of Penzias and Wilson and the theoretical explanations of Dicke, Peebles, Roll, and Wilkinson published in two successive papers in the "Astrophysical Journal" [3],[4].

The microwave background radiation data measured by Penzias and Wilson

---

[3] R. H. Dicke, et al., *Cosmic Black-Body Radiation*, Astrophysical Journal 142, 414 (1965).

[4] A. A. Penzias and R. H. Wilson *A Measurement of Excess Antenna Temperature at 4080 Mc/s*, Astrophysical Journal 142, 419 (1965).

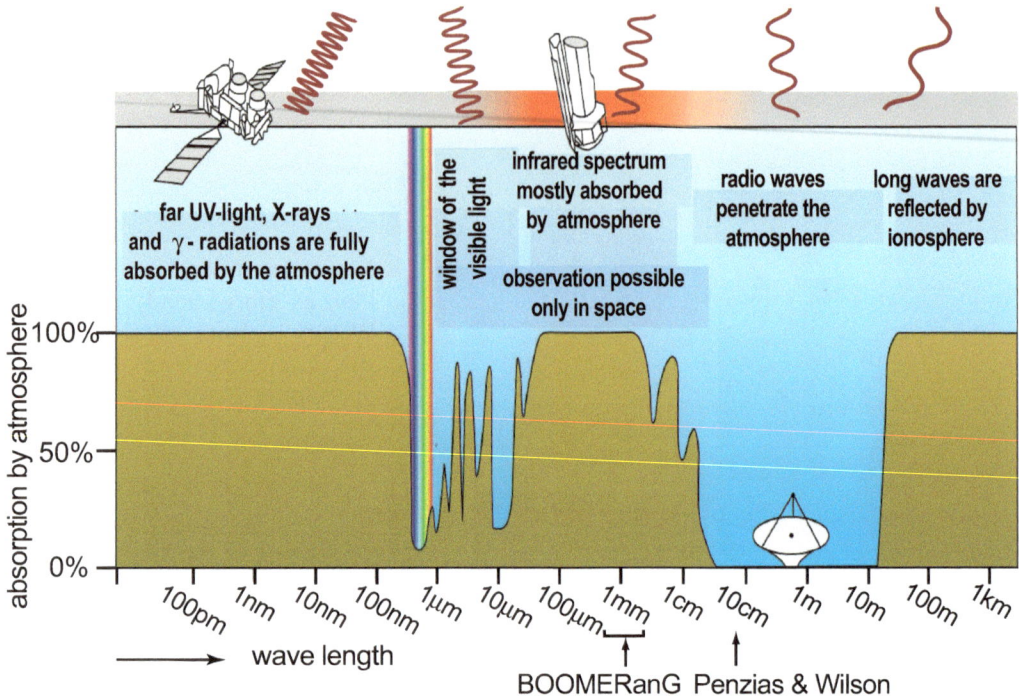

Fig. 8.3: Transmittance of the Earth's atmosphere to electromagnetic radiation. The arrows indicate the wavelength ranges of the Penzias/Wilson and the BOOMERanG experiments. The area highlighted in red in the upper part of the image indicates the wavelength range of the cosmic background radiation.

revealed for the Universe an equivalent temperature of $3.5 \pm 1\,\mathrm{K}$, an extraordinary result by any standards.

Penzias and Wilson received the Nobel Prize in Physics for their discovery in 1978.

The photo in Fig. 8.2, showing Arno Penzias and Robert Wilson in front of their 6 m horn antenna, is one of the most frequently shown photos in connection with the discovery of the background radiation.

Side note: The extraordinarily productive program using passive communication satellites such as ECHO was discontinued after about seven years and replaced in the course of further developments by programs using active satellites.

The discovery of the background radiation marked a turning point in cosmology and the beginning of intensive research activities, both with satellites, which have the advantage of being able to survey the entire celestial sphere, and with Earth-bound balloon telescopes, which, in contrast to satellites, represent a considerably less expensive alternative. However, balloon telescopes have to deal with absorption by the Earth's atmosphere (see Fig. 8.3) and with all kinds of interfering radiation from radar and radio waves as well as with the problem that the flight-path of the balloons cannot always be predicted.

On November 18, 1989, NASA launched the **Co**smic **B**ackground **E**xplorer satellite COBE for the first complete sky-wide survey of the background radiation. The scientific project management was in the hands of George Smoot (Lawrence Berkeley Laboratory, University of California) and John Mather (NASA), who received the Nobel Prize in 2006 for the measurement of the blackbody spectrum of the Universe and for the discovery of the anisotropies it contains.

The satellite was placed in a polar orbit at an altitude of about 900 km, so that the sensitive part FIRAS *(**F**ar **I**nfra**R**ed **A**solute **S**pectrophotometer)* was always facing away from the Sun and the Earth. COBE is still in that orbit today.

COBE successfully measured the cosmic microwave radiation over a wide spectral range and determined the temperature of the Universe to T=2.725 K with an absolute accuracy of 0.03% ($\approx 8 \times 10^{-4}$ K), showing that the temperature radiation follows Planck's radiation formula almost exactly.

In fact, the Universe is the most ideal "Blackbody" known so far.

As a big surprise it turned out that a spatial microstructure was imprinted in the radiation spectrum, which, after carefully subtracting all perturbation effects, remained in the spectrum as a property of the background radiation (Fig. 8.4). It quickly became clear that physics from the very early phases of the Universe was hidden in these fluctuations [5].

---

[5] D. J. Fixsen et al., *The Cosmic Microwave Background Spectrum from the Full COBE FIRAS Data Sets*, The Astrophysical Journal, 473:576 (1996).

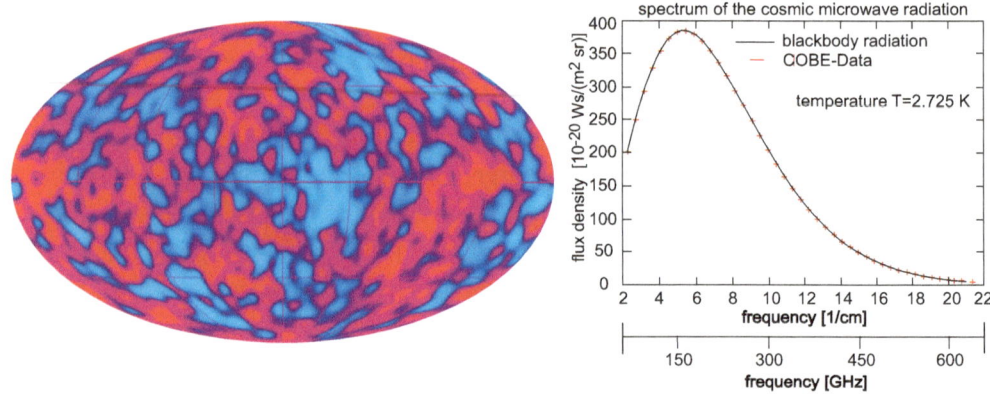

Fig. 8.4: Left: the temperature spectrum of the microwave background radiation measured by COBE. The temperature fluctuations (color-coded here) are in the order of 50 μK. **Right:** the measured frequency spectrum of microwave radiation compared to Planck's theory of blackbody radiation for a temperature of 2.725 K. Error bars are inflated by a factor of 10.

The next big knowledge leap came along with the BOOMERanG balloon telescope (**B**alloon **O**bservations **O**f **M**illimetric **E**xtragalactic **R**adiation **an**d **G**eophysics). BOOMERanG had two flights, in 1998 and 2003, both launched from McMurdo Station in Antarctica. The balloons rose to about 42,000 meters, thereby largely eliminating the atmospheric absorption of microwave radiation (see Fig. 8.3). At the same time, the circular polar wind at the South Pole carried the balloons back to the starting point after an approximately 2-weeks journey on the -79th latitude, which is also why the project got its special name. In contrast to COBE, only a small part ($\approx 3 - 4\%$) of the celestial sphere was visible to BOOMERanG that is, of course, the case for all Earth-based telescopes [6],[7].

Figure 8.5 shows the temperature map measured by BOOMERanG together with the embedded structures, which now at an angular resolution of 0.17° appear significantly more pronounced than those in the COBE experiment (see Fig. 8.4). Although this rather limited data set only allowed an early

[6] S. Masi et al., *The BOOMERanG experiment and the curvature of the Universe*, Progress in Particle and Nuclear Physics 48:243 (2002) and arXiv:astro-ph/0201137 (2002).

[7] C. B. Netterfield et al., *A measurement by BOOMERanG of multiple peaks in the angular power spectrum of the Cosmic Microwave Background*, The Astrophysical Journal, 571:604 (2002).

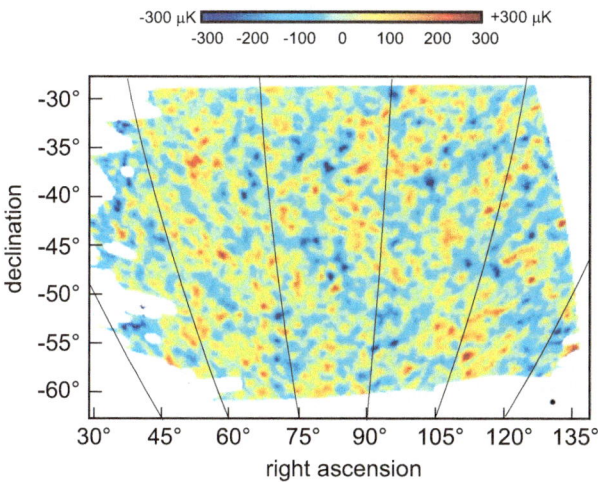

Fig. 8.5: Temperature data of the Universe from the BOOMERanG experiment for a section of the sky at the South Pole, shown in the equatorial polar coordinate system. The declination angle of $-90°$ would be the point directly above the South Pole. However, this point is not covered by the balloon on its circumpolar journey. The $15°$ separation lines of the right ascensions denote the rotation of the Earth within one hour.

and still rudimentary analysis, it already showed convincingly that properties and fingerprints from the early dynamical phase of the Universe were embedded in the temperature spectrum and that the temperature fluctuations were by no means purely statistical in nature. These findings should become even clearer and even more conclusive in the following satellite experiments and, thus put many alternative (and scientifically quite justifiable) "Big Bang" as well as "no-Big-Bang" theories to rest.

Almost concurrently with BOOMERanG, NASA launched in June 2001 the WMAP (**W**ilkinson **M**icrowave **A**nisotropy **P**robe) satellite with its 840 kg WMAP telescope, and in May 2009, the European Space Agency ESA launched the Planck satellite with its two-ton Planck telescope, each with the goal of making a more accurate and complete survey of the microwave background radiation.

Both satellite telescopes were parked near the Lagrange point L2 in the Earth-Sun system or put into orbit around this point. The Lagrange point L2 is located about 1.5 million kilometers away from Earth and is one of five Lagrange points existing in the Earth-Sun system (see Fig. 8.6). At point L2, one takes advantage of the Earth's shadow to protect the probes from intense solar radiation. Both probes left the Lagrange point at the end of the mission (WMAP in 2010 and Planck in 2013) and were placed in an orbit around the Sun that won't allow a collision with the Earth for any

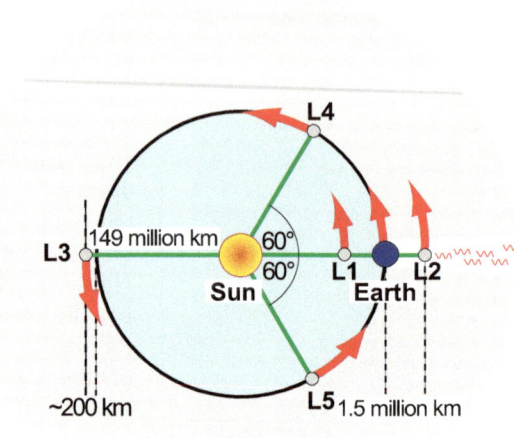

Fig. 8.6: Lagrange points L1 – L5 for the system Earth-Sun. All points have the same orbital period around the Sun, as if sitting on the axis of a wheel. The point L2 lies in the Earth's shadow of the Sun and sees the Earth only by night. The position is therefore particularly suitable for sensitive measurements of the cosmic background radiation. Several probes are currently located at this point, e.g., since September 2019 also the German-Russian Spectr-RG-Observatory with its X-ray telescopes eROSITA and ART-XC.

Spectr-RG – Spectrum, Röntgen, Gamma

eROSITA – extended ROentgen Survey with an Imaging Telescope Array

ART-XC – Astronomical Roentgen Telescope X-ray Concentrator

(Note added in print: In Jan-2022, also the James-Web telescope arrived at this point. Further, in Feb-2022, the eROSITA collaboration became temporarily inactive.)

time to come. This is a well-motivated safety measure since the Lagrange point L2, in contrast to L4 and L5, is an unstable residence point in the radial direction for any object. In the long term, i.e., perhaps after hundreds or thousands of years, either an uncontrolled crash to Earth will occur, or the object will move radially away from Earth. It is not possible to predict, which of these scenarios will occur or when — it is similar to placing a ball on top of a needle.

Figure 8.7 shows the temperature maps of the two probes in comparison. The agreement of the results from these two completely independent experiments is simply impressive. The only thing to consider here is that the temperature and angular resolution of the two experiments are different.

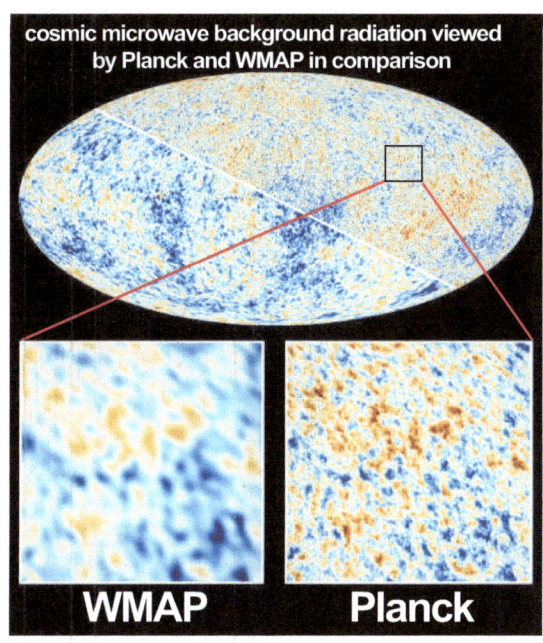

Fig. 8.7: Comparison of the temperature maps of Planck and WMAP. The angular resolution in the Planck experiment is about a factor of 2 better than in the WMAP experiment. Taking this difference into account, the detail enlargement shows an impressive agreement between the two experiments.

Furthermore, the amount of data from Planck is about 10 times as large as the one from WMAP. The values given by Planck and WMAP are[8),9)]:

Planck: $\Delta T \approx 10^{-6}$ K, $\Delta\Theta \lesssim 0.08°$ (at frequencies $\approx 1000\,\text{GHz}$)
data points: $\approx 1.43\,\text{million}$

WMAP: $\Delta T \approx 2 \cdot 10^{-5}$ K, $\Delta\Theta \lesssim 0.3°$
data points: $\approx 150{,}000$

In the following Fig. 8.8 the complete temperature map from the Planck experiment is shown. In this seemingly utterly confusing temperature chaos, the question arises: how can one extract quantitative information from this map, and how can it lead to robust conclusions? — Fortunately, students of natural sciences in the third or fourth semester already have the tools to give

[8)] Planck-Collaboration, *Planck 2018 results. I. Overview and the cosmological legacy of Planck*, arXiv:1807.06205v1 [astro-ph.CO] (2018), also publ. in Astronomy & Astrophysics 641, A1 (2020), and *Planck 2018 results. VI. Cosmological parameters*, arXiv:1807.06209v2 [astro-ph.CO] (2019). — All articles to *"Planck 2018 results"* are found as "Special Editions" in Astronomy & Astrophysics 641, A1 – A12 (2020).

[9)] G. Hinshaw et al., *Nine-year Wilkinson Microwave Anisotropy Probe (WMAP) Observations: Cosmological Parameter Results*, The Astrophysical Journal Supplement Series 208:19 (2013).

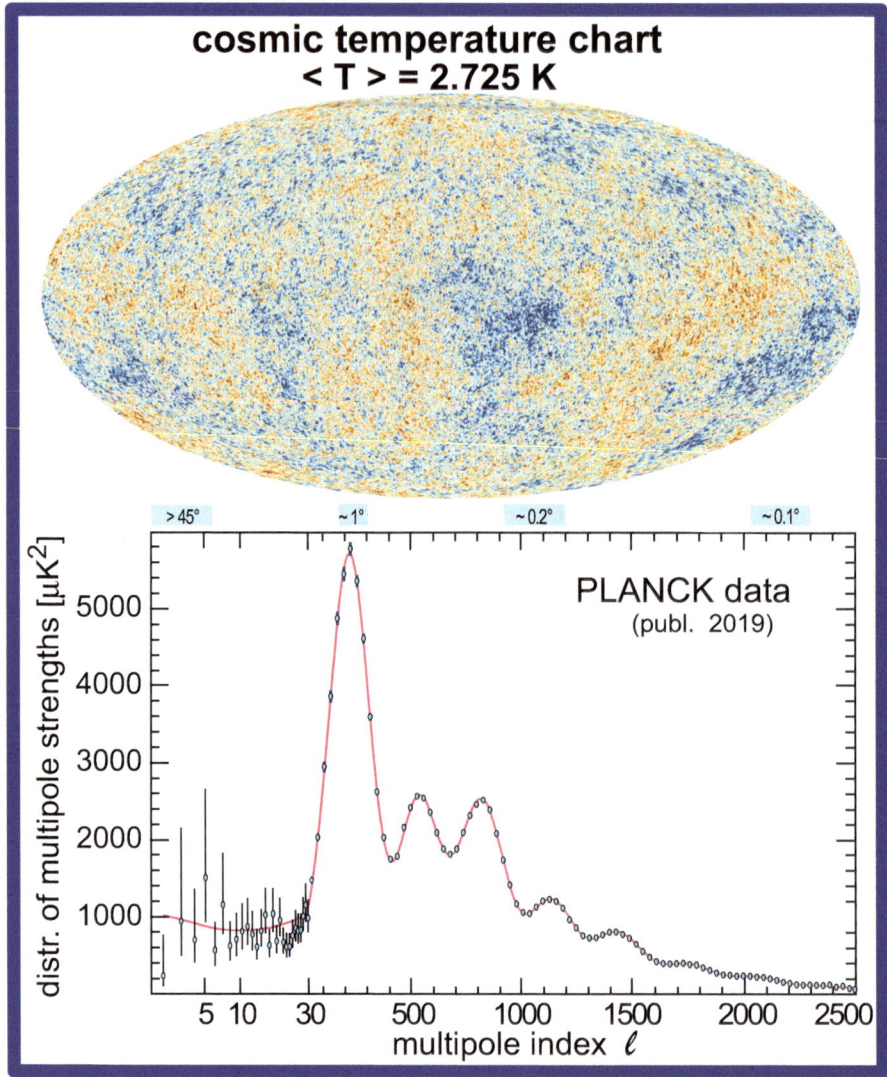

Fig. 8.8: **Top:** Cosmic temperature map from the Planck experiment for the whole celestial sphere. ($\Theta = [0°, 360°]$, $\Phi = [-90°, 90°]$). Temperature differences in the order of the temperature resolution $\Delta T \approx 10^{-6}$ K cannot be fully represented in this 2-dim form. **Bottom:** Multipole analysis showing size and frequency of the structures in from of a "power spectrum". It shows oscillations that are interpreted as acoustic reverberations of the "Big Bang". The height of the first maximum is directly related to the baryon density during the "Big Bang" epoch. The numbers at the top indicate the angular ranges, over which these structures (oscillations) are extended. — The data set includes about 1.43 million independent data points measured several times, from which the power spectrum was finally generated. This spectrum is described by the ΛCDM model (solid line) with considerable precision. Note: From $\ell = 30$ on, the scale changes from logarithmic to linear!
[Adapted from: Planck Collaboration, *Planck 2018 results. VI. cosmological parameters*, Astronomy & Astrophysics 641, A6 (2020)]

an answer to this question and perhaps to even perform an initial analysis on their own. The procedure to be used is called "multipole analysis" by means of "spherical harmonics". It is similar to the "Fourier analysis" of a frequency spectrum with various frequencies superimposed. Analytical methods of this type are used in almost all areas of the natural and engineering sciences, in radio and sound analyses, in various medical sciences, and even in economics and linguistics. Nowadays a large number of corresponding computing codes is freely available on the Internet.

Such an analysis performed for the cosmic temperature anisotropies yields as a result a so-called "power spectrum", shown in the lower part of Fig. 8.8. It correlates in strict mathematical formulation the size of contiguous structures (in units of the respective angular dimensions) with their abundances in the map — the larger the multipole order the smaller the structures. If the structures were purely statistical, the "power spectrum" would be flat.

This temperature analysis revealed a surprising result. On angular scales of about 1° one observes a pronounced oscillation maximum in the power spectrum. Interestingly, this scale corresponds approximately to the present size of those areas that were causally connected during the recombination phase (or the event horizon of the Universe at the time of 380,000 years). Since all these oscillations were uniformly imposed on all angular regions of this 1°-size in the temperature map, these regions must necessarily have been in causal contact at a much earlier time, e.g., in the pre-inflationary phase. They were eventually separated by inflation and re-enter our event horizon today (see Fig. 6.14), and since these areas all have exactly the same original history, they continue to resemble each other. Of course, also in future times more areas that were previously in causal contact, will enter future event horizons. The difference will only be that all oscillation maxima in the future power spectrum will shift to even smaller angles, they will further "jostle together" in the sky. But this process will not be recognizable in the course of a human life.

In a simplified picture, the observed oscillation maximum in the power spectrum is the fundamental frequency of the "sonic wave", which had been incorporated by the "Ba✳ng" of the "Big Bang" into the cosmic microwave radiation as a fingerprint of a fundamental baryonic oscillation, whose origin dates back to the pre-inflationary phase.

The temperature fluctuations are thus a direct legacy of the "Big Bang", and their distribution in the sky a direct consequence of inflation.

With the Planck probe one had a celestial telescope at hand that was capable of measuring oscillations down to the 8th harmonic (or 7 overtones), i.e., down to angular scales of about 0.08° (see Fig. 8.8). Particularly remarkable in the power spectrum of Fig. 8.8 is the fact that the frequencies of these overtones are in an almost integer ratio to the fundamental frequency. Music lovers know that in such a case the sound is especially harmonious!!

A theoretical description of the power spectrum shown in Fig. 8.8 succeeds by the so-called "**L**ambda-**C**old-**D**ark-**M**atter" (ΛCDM) model, which reproduces the data with an astounding precision. This model starts from relatively simple assumptions derived from the laws of hydrodynamics, thermodynamics, acoustics, and General Relativity, and is then applied to an expanding Universe. However, two new and hitherto unknown "entities" have become a central part of the ΛCDM model. They embody physics that has not yet been included in the standard physics of elementary particles, however, the evidence of their existence is compelling even though their properties remain largely obscure. These are:

▶ **Dark energy:** The phenomenon of "dark energy" was discovered in the late 1990s. The name was derived from the mysterious behavior that dark energy, unlike all other forms of energy, acts gravitationally repulsive, which then leads to an accelerated expansion of the Universe,[10]. In reference to Einstein's cosmological constant, it is abbreviated with the letter "Λ". However, Einstein's cosmological constant was originally introduced by him to hold the Universe together. In some respect, the cosmological constant reappears today, 100 years later, though in a completely new garment.

To what extent this "Λ-expansion" is a leftover from the inflationary age remains a question to be addressed by future research projects.

---

[10] For this discovery, Saul Perlmutter of Lawrence Berkeley National Laboratory, Brian P. Schmidt of the Australian National University, and Adam Riess of the University of California, Berkeley (now at Johns Hopkins University, Baltimore) received 2011 the Nobel Prize in Physics.

▶ **Dark matter:** That there must be something like dark matter was already suspected by Fritz Zwicky in the 1930s based on his observations of redshifts in galaxy clusters. Ninety years later, there is a plethora of evidence for dark matter, though to this day almost nothing is known about its structure, except that there must be a lot of it around, and that dark matter is subject to the «*gravitational*» and maybe as well to the «*weak*» interaction. Thus, dark matter produces no light, absorbs no light and scatters no light — dark matter is completely transparent and therefore just the "Dark Side of the Universe".

In the above mentioned ΛCDM model, an important property of dark matter is nonetheless prejudiced, namely: it ought to be "cold". This means that dark matter particles should be "heavy", or massive enough that they swim along in the early cosmic temperature bath and perhaps even thermalize there by virtue of the «*weak*» interaction. But during the decoupling phase they ought not to possess relativistic velocities anymore. In some ways they resemble neutrinos, except that they are many orders of magnitude (estimated by a factor of about $10^{10} - 10^{13}$) more massive and therefore manifestly slower. If this "cold property" is assumed, the filamentary structures of galaxies and clusters of galaxies can be reproduced in a convincing way by computer simulations. This implies that "cold" dark matter particles are extremely effective structure-forming particles and may provide an explanation why the first proto-galaxies appeared already a few hundred million years after the "Big Bang". Neutrinos, on the other hand, do not have this "cold property" because of their tiny mass and consequently because of their relativistic velocities at all times. They immediately iron out the smallest irregularities in the gravitational potentials and, thus prevent any noteworthy structure formations right at the beginning. However, the super-light axions, which were also mentioned at the beginning of this book, also remain as valid "cold" dark matter particles in this discussion.

The results with the most significant parameters for this book, which were communicated by the Planck collaboration (Ref. [8] page 109), are compiled below. They do not differ significantly from the published data from the WMAP mission, but have much lower error margins. It is noteworthy that our Universe consists of about 95% largely unexplained dark energy and dark matter, and just about 0.4% of luminous matter such as the Sun, the stars and the galaxies that surround us as visible objects in the skies in the day or at night. This should create a certain cosmic modesty.

The values determined by Planck:

$\Omega_{baryon} = 4.90 \pm 0.04\%$ density fraction of baryons, whereby "baryons" is synonymous with "normal matter", i.e., mostly hydrogen and helium

$\Omega_{visible} \simeq 0.4\%$ density fraction of luminous matter, i.e., Sun, stars, galaxies

$\Omega_{dark} = 26.1 \pm 0.3\%$ density fraction of dark matter

$\Omega_{Matter} = 31.1 \pm 0.6\%$ density fraction of both forms of matter

$\Omega_0 = (0.999 \pm 0.004)$ total density in units of critical density

$\Omega_{dark-E} = (68.9 \pm 0.6)\%$ dark energy density as missing fraction

$H = 67.7 \pm 0.4 \, \frac{km}{s \, Mpc}$ Hubble-constant

$N(\nu) = 2.99 \pm 0.11$ measured number of neutrino species: because of an integer number: $\Rightarrow N(\nu) = 3$

$\sum(m_\nu) < 0.12 \, eV$ sum of the masses of all 3 neutrino types

$k = 0.001 \pm 0.002$ curvature parameter consistent with a flat Universe

The first quantity in this list, Ω_{baryon}, deserves a separate consideration. We saw earlier that the "acoustic" maximum in the power spectrum of the temperature map is linked in a direct way to the density of baryons in the Early Universe, since the baryonic reverberation resulting from the "Big Bang" imprints the largest fingerprint onto the early microwave radiation.

The baryon density extracted from this maximum (i.e., the density of normal matter, in this case hydrogen and helium) is $\Omega_{baryon} = 4.90 \pm 0.04\%$ in units

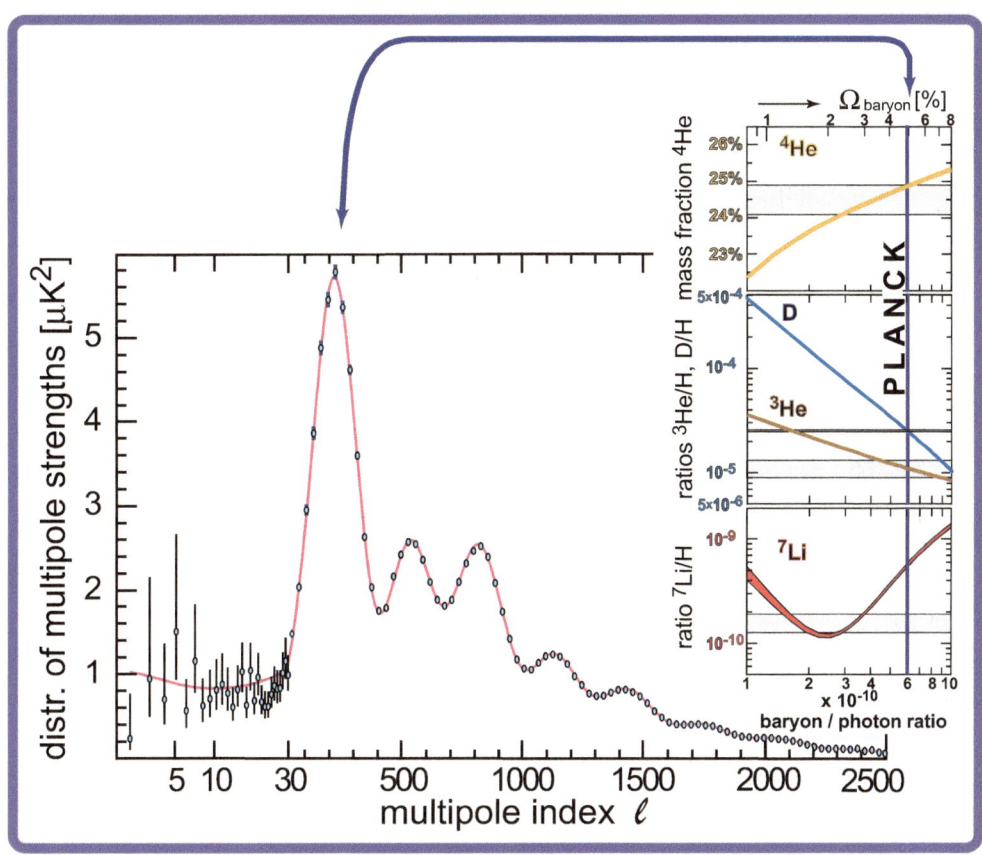

Fig. 8.9:   The microwave background radiation: an image of early nucleosynthesis —
380,000 !! years lie in between. Compare also Fig. 7.5 and Fig. 8.8.

of the Universe's total energy density. This value — although measured
today — reflects the value at a time 380,000 years after the "Big Bang".
Remarkably, this value is the same as the one extracted in a completely
independent way from the element abundances of early nucleosynthesis (see
Fig. 8.9 but also Figs. 7.5 and 8.8). However, nucleosynthesis took place in
the first 10 minutes !!

This is truly an extraordinary result, because it shows that the
laws of physics have not changed over this period from 10
minutes to 380,000 years after the "Big Bang".

Chapter 9

Late Nucleosynthesis and Supernovae

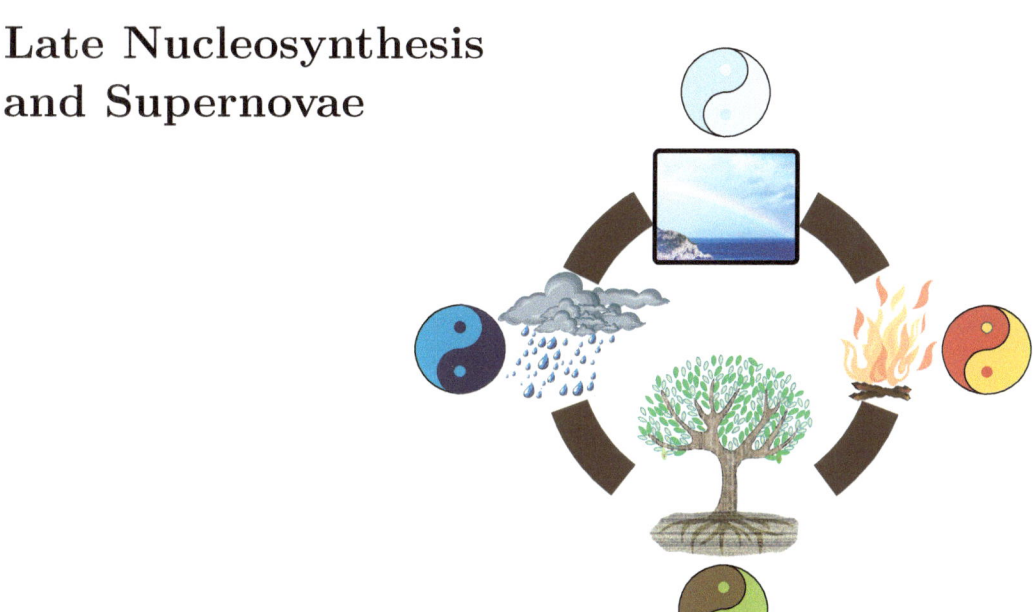

9.1   Elements, elemental questions

From early times until the Middle Ages, scientific life was simple — there were 4 elements only. This was sufficient for the description of life and for most of the events occurring. For alchemists, though, especially for those who wanted to make gold but didn't manage, this classification seemed at least questionable very early on. Finally, with the beginning of the industrial revolution, a general basic knowledge about the properties of substances and materials became more and more important, and slowly the realization grew that certain substances were obviously "elemental".

D. Frekers, P. Biermann, *Universe, Neutrinos, Stars and Life*, https://doi.org/10.1007/978-3-662-70729-6_9

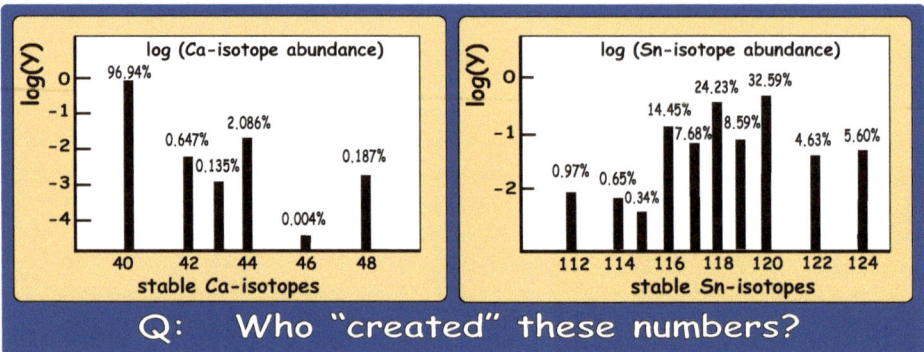

Fig. 9.1: Isotopic abundances of calcium (Ca) and tin (Sn). Calcium has 6 stable isotopes and tin a total of 10, which also makes this element the one with the most stable isotopes in the periodic table.

In the middle of the 19th century, the Russian chemist **Dimitri Mendeleev** and the German physician and chemist **Lothar Meyer** succeeded in a painstaking effort to establish a largely correct classification of these "elements". Both are considered today the founders of the periodic table of elements. And finally, the English chemist and physicist **Francis William Aston** is regarded as the discoverer of the isotopes. He succeeded in identifying by mass spectrometry a total of 200 of the nearly 300 naturally occurring isotopes[1], for which he was awarded the Nobel Prize in chemistry in 1922.

By looking backwards from today's perspective, these early developments give the impression that the physicists were rather slow in getting interested in the stable elements — and indeed, the unstable, radioactive elements were initially of much greater interest to them. It was not until the 1950s (!!) that they gradually began to ask simple and supposedly trivial questions,

---

[1] Note: Not all naturally occurring isotopes are stable. The most abundant natural radioisotope is potassium-40 (^{40}K, half-life 1.25×10^9 years). In addition, there are about 20 others, including a number of rare-earth metals. Furthermore, the decays from natural fission products of uranium and thorium occur in larger concentrations in the biosphere (the most important ones are: radon, radium, cesium, and technetium). Cosmic rays also produce radioisotopes that enter the biosphere in small concentrations (the most important: tritium (^{3}H, half-life 12.3 years), carbon-14 (^{14}C, half-life 5730 years), beryllium-10 (^{10}Be, half-life 1.6 million years), argon-39 (^{39}Ar, half-life 269 years), calcium-41 (^{41}Ca, half-life 103,000 years). Another and quite significant amount of natural radioisotopes resides in the interior of the Earth. These radioisotopes get injected into the biosphere by volcanism.

also to find that there were no answers to them, e.g.:

▶ Where do all these elements come from, where are the cauldrons in the cosmos?
▶ Why are some elements so rare (e.g., gold or lithium) and others so common (e.g., carbon or iron)?
▶ Why are all stable isotopes actually present in Nature, why has none been "forgotten" during production?
And above all:
▶ Why is the relative abundance of isotopes so seemingly completely arbitrary; who is the custodian of these numbers?

In Fig. 9.1 the last point from this sequence is exemplified for two different elements, i.e., calcium and tin. Calcium has 6 stable isotopes, of which calcium-40 (^{40}Ca) is the most abundant at 97% and calcium-46 (^{46}Ca) is the least abundant stable isotope at 0.004% — in fact, it is the rarest among all stable element isotopes. Tin is the element with the largest number of stable isotopes (10 in total), most of them have relative abundances in the single-digit percentage range and none of them is excessively rare.

The other questions from the above sequence can be articulated in an even more revealing way by Fig. 9.2. It shows the relative elemental abundances of all chemical elements and their isotopes as they are found both inside and outside our solar system [2]. Noteworthy is the still low abundance of lithium, which has hardly changed since the "Big Bang". Furthermore, the area of the rare-earth elements is specially marked to designate their extremely low abundance (partly even rarer than gold), a fact that also led to their naming.

---

[2] These element abundances have been determined over the last $40-50$ years with ever increasing accuracy. Earth rock, moon rock, Mars rock, meteorite rock and meteorite dust from the near and distant solar environment as well as extrasolar meteorite material were used as sample substances. Also, spectral analyses of stars provide important information about elemental abundances.

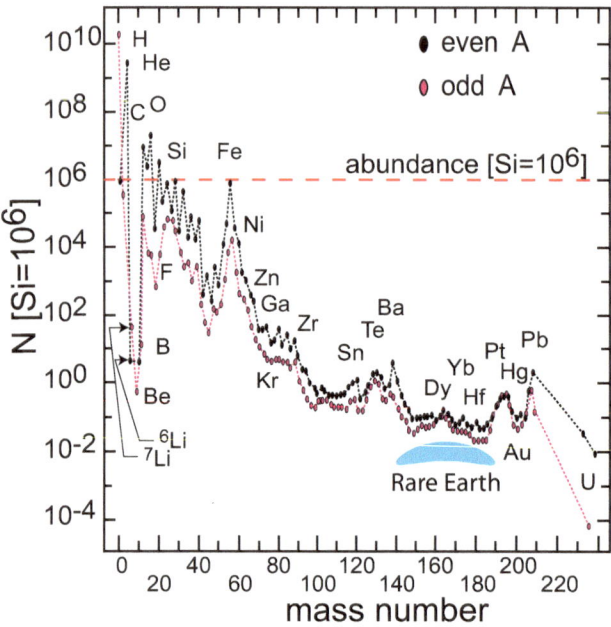

Fig. 9.2: Relative solar element abundances normalized to the element abundance of silicon (arbitrarily set to 10^6). The abundances for even and odd mass numbers are shown separately. Clearly recognizable is the predominance of iron and the elements close to iron. Remarkable also the peak abundances (approx. a factor of $50-100$) in the tellurium/barium and the lead region.

9.2   The Sun — one among many

It may be useful to start by becoming familiar with the properties of our central star, the Sun, and the remarkable physical processes that so directly and imminently pertain to our own existence.

Its fact sheet

The Sun is a normal, inconspicuous, maybe even a somewhat undersized star in the Orion arm of our Milky Way. In diameter it is about 100 times larger than the Earth (about 1.4 million kilometers) and has a mass of about 2×10^{30} kg (in comparison: mass of the Earth about 6×10^{24} kg). It brings about 99.9% of the total mass of the solar system on the balance. Its mean density is quite modest at about 1.4 g/cm^3, yet its enormous mass results in a gravitational force on the surface of the Sun, which is about 28 times that on Earth (i.e., $28 \cdot g_{\text{Earth}}$). Its "outside temperature" is about 5800 K, but inside there is a temperature of about 15 million K, which is maintained by the energy output during the fusion of hydrogen into helium.

Fig. 9.3:   An image of the Sun (here in the far ultraviolet range, 30.4 nm) with the planets in relative size comparison. The smallest planets Mercury and Mars are barely visible in this image.

Converted into electron-volts, this temperature is just 1.3 keV. The density in the inner solar region reaches 160 g/cm³, about 14 times the density of lead, and is stabilized at this value through the interaction of attractive gravity and the outward facing pressure of the gas and the radiation. The vast majority of the Sun's mass is hydrogen and helium, namely about $(67.8 \pm 3.5)\%$ and $(30.6 \pm 1.5)\%$ [3]; this also means, the Sun still bears the signature of an elemental composition derived from early nucleosynthesis in the first 10 minutes (see chapter 7). The remaining $\approx 1.5\%$ of the solar mass consists of heavy elements (A > 4), suggesting that the solar material has already been recycled several times from the material of earlier stellar explosions.

---

[3] The percentages given here come from a recent, quite involved theoretical solar model calculation by A. Noels-Grötsch and N. Grevesse that is not yet published. The two authors have kindly allowed us to quote their calculated values in this book.

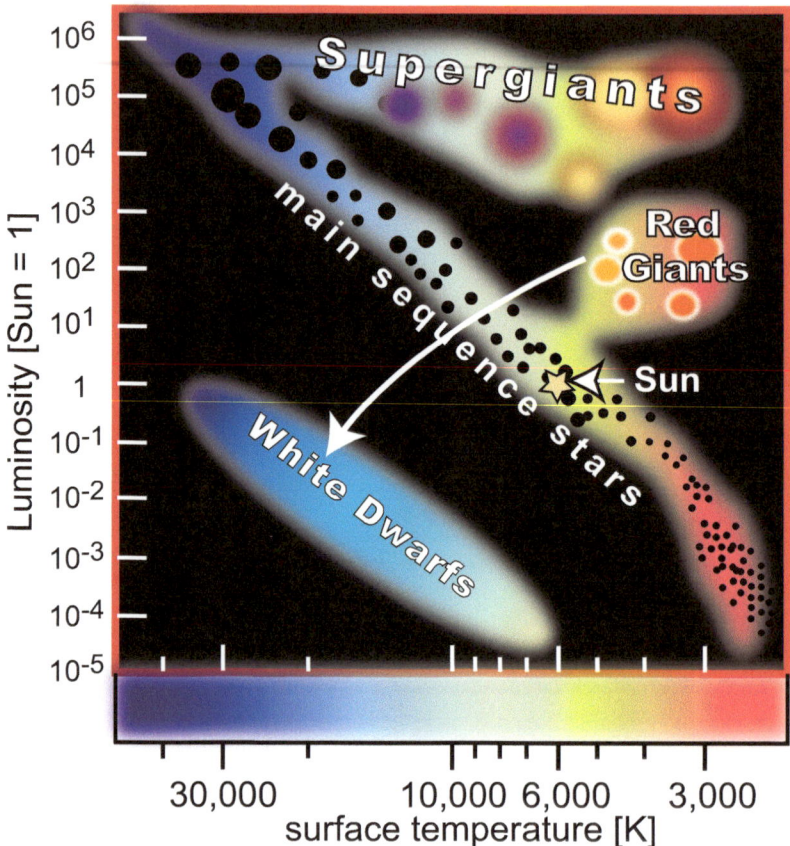

Fig. 9.4: Schematic representation of the stars in the Hertzsprung-Russell diagram.

The individual stages of stellar evolution that all stars, including the Sun sooner or later have to pass through, are shown in an impressive form in the Hertzsprung-Russell diagram [4] (Fig. 9.4). For more than 100 years now, it has been successively extended by continuous astronomical observations. In the diagram the color (i.e., temperature) of a star is correlated with its luminosity. Obviously, stars are subject to an order, the stellar parameters are not arbitrary — stars have no freedom!

More than 90% of all stars are arranged in the diagram on the so-called main sequence, where they, like the Sun, fuse hydrogen to helium by nuclear

---

[4] E. Hertzsprung (1873 – 1967) Danish astronomer; Henry Russell (1877 – 1957), American astronomer.

fusion. Also like the Sun, they remain in a stable equilibrium between gravity and radiation pressure for several billion years. This in turn demands that particularly massive main-sequence stars must exhibit a high luminosity (thus, high radiation pressure), as this is the only way they escape collapse.

At the end of their evolution, stars convert within a short time (therefore the bottleneck in the diagram appears as a relatively empty area) to Red Giants or to Supergiants. They then perish, depending on their initial size, either as White Dwarfs or explosively as neutron stars (in extreme cases even as Black Holes). This final and inescapable process lasts between 100 and 600 million years.

In the Hertzsprung-Russell diagram, the Sun is located in the lower half of the main sequence in a relatively small region and in good company with most of the stars in the Universe.

Its fuel and its fuel consumption

The energy of the Sun is released in its interior by the fusion of 4 protons to a helium nucleus. The fusion takes place in the inner quarter of the Sun above a temperature of about 10 million K and above a density of about $100\,\text{g/cm}^3$. The nuclear physics process is schematically shown in Fig. 9.5. It starts from two different branches with a total of 6 protons resulting in a single ^{4}He plus 2 protons left behind, which are returned to the thermal bath to be available for further fusion. Two neutrinos leave. This initial process of energy production, called the *pp*-(I) cycle, is the same for all stars and is quite different from the process of early, post-"Big Bang" nucleosynthesis. Further, to avoid misunderstanding, it is not the process that one would consider in any planned terrestrial fusion reactor.

The *pp*-cycle has some noteworthy features, which give reason for a closer look. It consists of a total of 4 subcycles *pp*-(I) to *pp*-(IV), of which the two most important ones, *pp*-(I) and *pp*-(II) with their sequences (I-1), (I-2), (I-3) and (II-1), (II-2), (II-3), respectively, are briefly described here:

$$\underline{pp\text{-(I-1)}:}\quad p + p \longrightarrow d + e^+ + \nu$$

This process is initiated by the «*weak*» interaction, already recognizable by the fact that a neutrino is involved. For this very reason, the reaction rate is

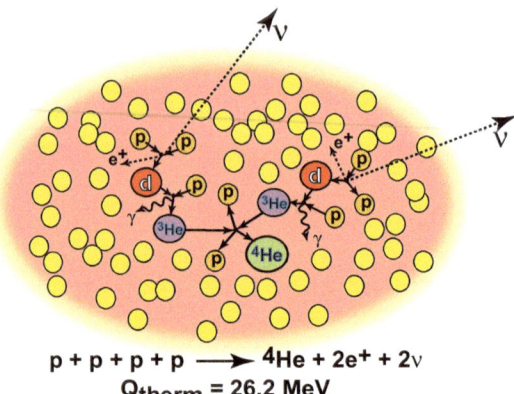

$$p + p + p + p \longrightarrow \text{4He} + 2e^+ + 2\nu$$
$$Q_{therm} = 26.2 \text{ MeV}$$

Fig. 9.5: Schematic representation of the primary proton-proton cycle in the Sun with ^{4}He as the final product. The reaction $p + p + p + p \longrightarrow ^4$He occurs via the intermediate steps $2\times(p+p \to d)$, $2\times(d+p \to ^3$He), and ^{3}He $+^3$He $\to^4$He $+2p$. With each cycle, 2 neutrinos escape. A total of about 26.2 MeV finally remains as thermal energy in the Sun.

extremely small, and the fact that the electrostatic repulsion potential keeps both protons at a distance does the rest to suppress the reaction rate yet again by several orders of magnitude[5]. Thus, it has not yet been possible to reproduce this reaction experimentally under laboratory conditions, nor will this be possible in the foreseeable future. Fortunately, the underlying physics is well known, so that a theoretical calculation of the rate under the conditions prevailing in the Sun is possible with high accuracy. According to this, the average lifetime of a proton in the central region of the Sun is about 9 billion years, before finally a collision process according to pp-(I-1) successfully leads to the fusion of a deuteron[6].

- **This extremely long time ensures us the long life of the Sun !**

---

[5] The electrostatic repulsive potential has a magnitude of about 1.44 MeV. At a temperature of 15 million K, the average kinetic energy of the protons is about 1.9 keV $(= \frac{3}{2}k_BT)$ and the minimum distance of 2 protons would be about 750 times the proton radius $(\approx 750 \times 10^{-15}$ m). Only the high energy part of the thermal energy distribution and the quantum tunnel effect can occasionally bring the protons close enough so that the reaction according to pp-(I-1) succeeds.

[6] To avoid misunderstanding, here a comparison with the lottery: The chance of winning the 6-draw lottery with a certain combination of numbers is 1 in 14 million, i.e., it takes an average waiting time of 7 million weeks (with 2 draws per week) to hit the right combination of numbers. Nevertheless, a main prize is determined in almost every drawing, since the number of tickets submitted (i.e., the number of trials) is of the same order of magnitude.

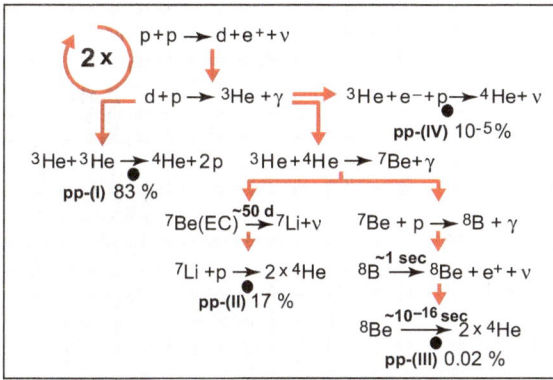

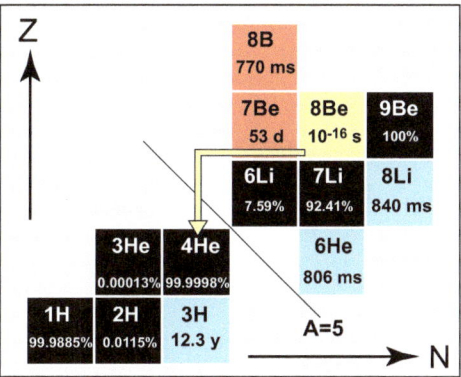

Fig. 9.6: Sequence of the various proton-proton cycles in the Sun with ^{4}He as final product. With each cycle 2 neutrinos escape. In total, about 26.2 MeV remain as thermal energy in the Sun.

Fig. 9.7: Section of the table of isotopes. Elements and isotopes are arranged as a function of neutron and proton number (atomic number) N, Z. Stable isotopes are shown in black, β^- and β^+ emitters in blue and red. ^{8}Be (yellow box) decays into two ^{4}He nuclei.

$$\underline{pp\text{-(I-2)}:} \quad d + p \longrightarrow {}^3\text{He} + \gamma$$

This process is mediated by the «*strong*» interaction and is therefore extremely "fast". The average lifetime of a deuteron in the core region of the Sun is just 4 seconds. During this time it is not possible for the deuteron to find another deuteron to initiate, for example, the equally "fast" reaction $d + d \longrightarrow {}^3\text{He} + n$, i.e., with a neutron as the final product. Neutron production would be highly toxic to the Sun — the Sun would explode in a runaway reaction in a very short time.

• **There was a fortuitous choice of interactions in the Early Universe.**

$$\underline{pp\text{-(I-3)}:} \quad {}^3\text{He} + {}^3\text{He} \longrightarrow {}^4\text{He} + 2p$$

This process is also mediated by the «*strong*» interaction, but is considerably suppressed owing to the $3-4$ times stronger electrostatic repulsion. The average lifetime of the ^{3}He is about 1 million years. During this time, the interior of the Sun accumulates enough ^{3}He so that the above reaction with two ^{3}He can finally be set in motion, thus establishing a balance between ^{3}He production and ^{3}He destruction.

With the production of sufficient ^{3}He, a concurrent branch into the *pp*-(II) cycle opens up. In this, ^{3}He fuses with the highly abundant ^{4}He in the Sun to form ^{7}Be:

pp-(II-1): ^{3}He $+ ^4$He $\longrightarrow ^7$Be $+ \gamma$

Beryllium-7 (^{7}Be) is unstable and converts to ^{7}Li plus a neutrino after a half-life of just 50 days through capture of an electron ("electron capture", EC), whereby the neutrino carries $862\,$keV energy out of the Sun:

pp-(II-2): ^{7}Be $+ e^- \longrightarrow ^7$Li $+ \nu$

The average residence time of ^{7}Li is a few hundred years. In the course of this, ^{7}Li is converted back ("burned up") to two ^{4}He by virtue of the reaction:

pp-(II-3): ^{7}Li $+ p \longrightarrow 2\ ^4$He

One may easily verify that this cycle creates one ^{4}He and one neutrino from a ^{3}He, a proton and an electron, whereby three protons were already consumed in pp-(I) to synthesize ^{3}He with the release of one more neutrino and one positron. Note: With the exception of neutrinos, everything else remains confined inside the Sun.

The pp-(I)- and pp-(II) cycles are the two main cycles, in which four protons fuse to form one helium-4 (i.e., $4\,p \rightarrow\ ^4$He), whereby the pp-(I) cycle contributes 83.3% and the pp-(II) cycle 16.68% to the total reaction. The remaining 0.02% proceed through the pp-(III) and pp-(IV) cycles, which can be seen from Fig. 9.6. In all these cycles, there are four protons at the beginning and one ^{4}He and two neutrinos at the end. Stable nucleosynthesis beyond mass A = 4 is not possible in the Sun during this period of "hydrogen burning", again because there is no stable element with mass A = 5 (see Fig. 9.7).

- **The "ashes" of the solar burning cycle are thus invariably ^{4}He.**

All cycles taken together convert on average about $26.2\,$MeV into thermal energy inside the Sun, and in each of them exactly 2 neutrinos are produced. They leave the Sun without interaction and take away an average energy of about 0.5 MeV. These "solar neutrinos" can therefore provide direct information about the interior of the Sun — of course, only if they are being captured (i.e., detected).

The consumption of the hydrogen in the Sun can now easily be calculated. The radiation power of the Sun measured on Earth, i.e., at a distance of 149.6 million km is $1.367\,\text{kW/m}^2$. Therefore, to generate this power, the Sun must convert about 610 million tons of hydrogen into helium per second. This sounds truly astronomical! However, with the high density inside the Sun, this is just about the volume of a sphere with radius of 100 m, a miniscule size compared to the dimensions of the Sun.

• **The Sun still has enough fuel for the next 5 – 6 billion years.**

Its life expectancy and destiny

The present age of the Sun is about 4.6 billion years — the Sun is therefore a comparatively young star. Its life expectancy is estimated to be another 5 – 6 billion years, of course, if nothing unexpected intervenes, e.g., like a nearby supernova explosion or the collision with our neighbor galaxy Andromeda, which is bound to happen in about 3 – 4 billion years. Any such event will surely modify the scenario drawn up in the following (see Fig. 9.8).

The hydrogen supply for the energy production in the Sun is finite, and the change of the solar activity, which will set in ever so slowly within the next 100 – 200 million years, will irrefutably affect all inner planets, including planet Earth. This development will continue and will increasingly get worse, so that already long before the Sun's termination date the planetary system in its present form will cease to exist.

In the core of the Sun the helium-"ashes" created by the hydrogen fusion process keep accumulating relentlessly, and in order to absorb the progressive gravitational contraction in order to maintain stability, temperature and radiation pressure must increase. This can only be achieved by successively increasing hydrogen fusion and consequently by increasing hydrogen consumption. An increase in energy production very effectively raises the temperature (due to the change in entropy) to the 20th power, i.e., $\Delta T \propto \Delta E^{20}$. The process makes the outer solar radius grow, first slowly though, but after about 4 – 5 billion years from now, the expansion accelerates. A hydrogen-fusion shell forms around the central helium region, which leads to a much more efficient energy production. As more and more

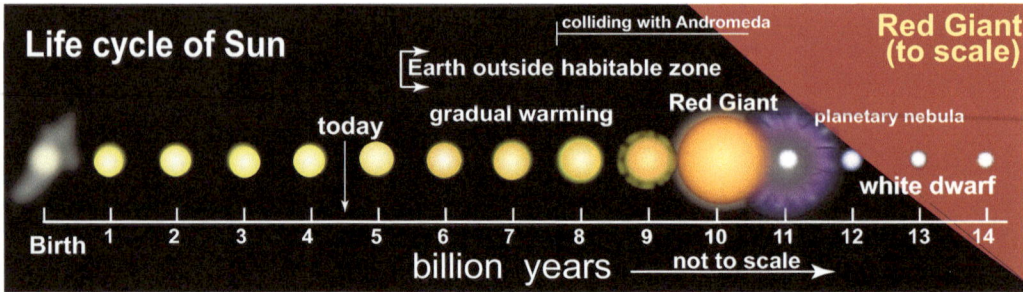

Fig. 9.8:   Life cycle of the Sun. Indicated is the approximate true to scale size of a Red Giant. A White Dwarf true to scale would be about 100 times smaller than indicated.

helium is being produced and falling into the central region, the hydrogen shell also keeps growing steadily as a result of growing gravitational compression. The luminosity increases $40-50$ fold and the solar radius about 10 fold. With the surface being accordingly enlarged about 100 fold, the Sun appears more and more in the red and longer wavelength region, i.e., the Sun enters the transitional region of turning into a Red Giant. Within less than a billion years it moves to the upper right in the Hertzsprung-Russell diagram of Fig. 9.4. By the growing loss of mass (about 10,000 times the present value) and the intense solar wind associated with it, the inner planets Mercury and Venus will initially be lifted to safer higher orbital radii, but safety won't last for long.

For Life on Earth things had already turned bad much, much earlier. In fact, within only a few 100 million years from now, living on planet Earth starts to get uncomfortably warm, and after about another 500 million years, Earth's Global Temperature will have risen to about $30°$ C (presently at $15°$ C). Earth dries up. And yet another $500-600$ million years later, planet Earth will sure find itself outside the habitable zone (cf. Fig. 9.8). From a cosmic perspective the "Earth-and-Life" intermezzo was all in all uneventful and inconspicuous.

In the course of the Sun's final 500 million years, i.e., starting at about $5-6$ billion years from now, the situation for the Sun also gets to be increasingly apocalyptic. The density of the hydrogen shell around an increasingly dense helium core has reached such critical proportions that the Sun emits 1500 times the present radiation, thereby expanding to about 150 million kilometers. The inner planets Mercury and Venus have been swallowed up long ago, and the solar surface now reaches up to Earth, which was initially also pushed up onto a higher orbit. The density of the Sun's outer surface region near Earth is small, only about $10^{-7}\,\mathrm{g/cm^3}$, yet the temperature of the gas is about $3000-3500$ degrees. More and increasingly faster helium is being produced in the core region of the Sun, where the temperature has grown to $100-150$ million degrees and the density risen to almost $1000\,\mathrm{kg/cm^3}$. The helium core enters a state of "degenerate rigidity", which is a condition dictated by quantum mechanics. Pressure and temperature no longer change the spatial expansion because the electrons, having the property of fermions, successfully resist further contraction. They cannot be confined to a smaller space.

Under these conditions a new (and extremely effective) energy source opens up, which is now located in the very interior of the Sun. It comes to "helium burning" via the so-called triple-α process (see next setion). In stars the size of the Sun, the process starts suddenly and explosively with violent bursts of so-called "helium flashes". The first main and most violent flash reaches a luminosity that is about 10^{10} times that of the present Sun for few seconds. These "helium flashes" are recurring every few 100,000 years until they eventually die down after about 1.5 million years. They abruptly lift the degeneracy allowing a more effective cooling of the star, and lead the star into a relatively quiet helium burning phase as a Red Giant, which from now on produces carbon via the triple-α process and oxygen via fusion of carbon and helium, both at about equal rates.

After another $50-100$ million years in this phase, the Sun perishes unspectacularly as a "White Dwarf". The mass of the Sun is simply not sufficient to ignite the next fusion process after the triple-α process. White Dwarfs are comparable in size to Earth, but have masses in the order of 60% of the former star. They appear in the lower left part of the Hertzsprung-Russell diagram (see Fig. 9.4). Their gravitational acceleration on the surface is about $(300{,}000-500{,}000) \cdot g_\mathrm{Earth}$, according to the star-mass/Earth-mass compression ratio. The White Dwarf closest to the solar system is the companion of Sirius, 8.611 light-years away. It has the properties here described.

White Dwarfs are by no means rare objects. The Gaia satellite of the European Space Agency has recently identified about 14,000 White Dwarfs within a 326 light-year radius of Earth. Of course, this has created an enormous activity to study these objects in detail (see e.g., Ref. [7])

In the end, a White Dwarf simply cools down. This is an extraordinarily slow process that can easily take $10^{10} - 10^{15}$ years. Eventually (in case there is still anybody around) this object escapes further observation as a "Brown Dwarf" or "Black Dwarf". Such an invisible "black" structure consists mainly of crystalline oxygen and carbon with a central density of about $1,000 - 10,000 \, \text{kg/cm}^3$ ($100,000 - 1$ million times that of lead). Thus, for carbon, perhaps even a giant diamond could be formed in the end — however, not to arouse any special desires: this statement is speculative.

9.3   Fascination triple-alpha

Carbon and oxygen are the two most important elements of Life, and after hydrogen and helium also the most abundant elements both in the solar system and in the whole Universe. This fact was already known in the 1950s, but there was no conclusive explanation as to how the formation mechanism of these two elements in such large quantities should take place. It was also clear that a triple-α process offered the only possibility to jump over the gap at mass number $A = 5$ and $A = 8$ in stellar nucleosynthesis, but all calculations showed that the rates for a fusion of three ^{4}He nuclei to ^{12}C with a subsequent fusion of another helium nucleus to ^{16}O, let alone the rates for a straight 4-body fusion to ^{16}O (i.e., $4 \times^4 \text{He} \rightarrow ^{16}\text{O}$) were by many orders of magnitude too small. It was obvious that the nucleosynthesis to produce carbon and oxygen using the triple-α process needed to be re-thought — but what were the things that had been overlooked so far? The magnitude of this problem was not to be underestimated because the synthesis of carbon and oxygen, in addition to being central to the existence of Life, it is also a basic pre-condition for a successful synthesis of all other elements in the further course of stellar evolution. If there is no carbon and no oxygen, there isn't anything else either.

The idea of a "thermal resonance" in the triple-α process, which goes back to Fred Hoyle (Fig. 9.10), finally led to the decisive breakthrough. It

---

[7] M. Kilic, et al, *Gaia reveals evidence for merged white dwarfs*, Monthly Notices of the Royal Astronomical Society: Letters 497, 113 (2018).

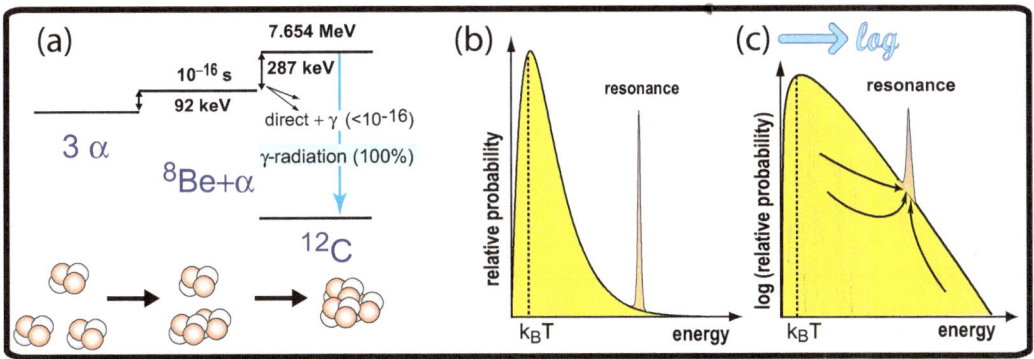

Fig. 9.9: Illustration of the triple-α process and the thermal resonance. In (a) two α-particles first combine to a ^{8}Be, and then a third α-particle fuses to ^{12}C by virtue of a resonance at 7.654 MeV in ^{12}C. This resonance de-excites by a 100% γ-transition to the ^{12}C ground-state. In a simple conceptual representation, indicated on a linear scale in (b) and on a logarithmic scale in (c), this resonance "eats" a hole into the thermal energy spectrum of the α-particles. However, thermalization immediately fills it again, so that fusion continues unhindered at the resonance position, and only there. The figure is for illustration and not to scale.

was initially heavily criticized as a conjuring trick and rejected with the argument that too many and too precisely coordinated coincidences were required, and therefore an "Intelligent Design" had to be brought into play. Even after the experimental verification, a certain magic around the "triple-α process" has persisted, because:

All stars, regardless of size, must pass through this needle eye of coincidences, which is so narrow and so finely tuned towards the synthesis of the life-essential carbon and oxygen elements that it leaves every astrophysicist and nuclear physicist in awe.

The triple-α process proceeds as before via the fusion of three helium nuclei (α-particles) into one carbon-12 (^{12}C) nucleus, but two decisive features come into play, each of which alone has little to no effect and only in concert produce an effective fusion probability. The various steps of reasoning are illustrated in Fig. 9.9 and shall be outlined in the following.

In the first stage, a ^{8}Be nucleus is formed from two α-particles. Energetically, though, ^{8}Be lies 92 keV above the energy level of two α-particles. The two α-particles must therefore provide this minimum energy first, in order to get a chance at all to form ^{8}Be. However, a stellar temperature of 100 million degrees corresponds to just 8.7 keV, and thereby only α-particles lying in the high-energy part of the energy distribution are able to initiate the reaction. A by far more unfavorable condition is, that ^{8}Be, once formed, decays back to two α-particles already after 10^{-16} seconds. While this time is a factor of 10^5 longer than the simple "fly-by" of two α-particles, it is clearly much too short to capture another α-particle in a subsequent reaction to form ^{12}C. The two components of ^{8}Be are returned to the thermal bath, and the carbon/oxygen problem remains unsolved.

Fred Hoyle then realized that a special mechanism was needed to accelerate the reaction. He postulated that there ought to be a state in the excitation spectrum of the ^{12}C nucleus with exactly the "correct properties" to produce a strong resonance for the fusion of an α-particle with this extremely short-lived ^{8}Be [8]. This state in the ^{12}C nucleus would have to be energetically in the range of the thermal energy spectrum of the α-particles as indicated in part (a) of Fig. 9.9, so that an α-particle with exactly the resonance energy can be removed from the thermal spectrum and fuse with the

Fig. 9.10: Fred Hoyle, British astronomer and mathematician (1915 – 2001). The state in ^{12}C at 7.654 MeV is named Hoyle state after him.

^{8}Be to form a ^{12}C nucleus. Also, another critical requirement in the long list of "correct properties" demanded that this resonant state, once formed, makes a transition to the ground state via γ-ray emission with a 100% probability, a condition considered utterly unlikely.

In the perception of the triple-α process, the resonance "eats a hole" into the thermal α-particle spectrum. Thermalization re-fills this "hole" instantaneously, and thus provides a constant re-supply for further fusion processes at exactly this resonance energy, and only there.

When assuming such a process, it is from thereon a relatively easy task to determine the parameters of this postulated resonance, like resonance energy

---

[8] F. Hoyle, *On Nuclear Reactions Occurring in Very Hot Stars I. The Synthesis of Elements from Carbon to Nickel*, Astrophysical Journal Supplement 1, 121 (1954).

and resonance width. One simply takes the observed carbon abundance in the Universe as a fixed input. Fred Hoyle (see photo Fig. 9.10) calculated that this resonance should be located at about 7.68 MeV in ^{12}C, which is about 300 keV above the ^{8}Be+α threshold (see part (b) and (c) in Fig. 9.9). And in fact, shortly thereafter this state was experimentally verified at 7.654 MeV (i.e., 287 keV above the ^{8}Be+α threshold) with all the predicted "correct properties".

Fred Hoyle's idea of a "thermal resonance" was thus experimentally confirmed in a grandiose way. However, he was not awarded a Nobel Prize. In order to find out, which motives possibly played a role here, the reader is referred to Fred Hoyle's biography. Nevertheless, the name Fred Hoyle is much more widely known among astrophysicists and nuclear physicists than that of many "regular" Nobel Prize laureates.

The fusion gamble takes another interesting course, because the fusion of helium and carbon to oxygen (^{12}C+$\alpha \rightarrow ^{16}$O+γ) also follows a similar resonance pattern. One might now assume that this situation continues in the next step to neon-20 (^{16}O+$\alpha \rightarrow$ ^{20}Ne+γ), but neon-20 "blocks". Neon-20 does not have a "fitting" state for such fusion and is synthesized only in the explosive scenario of a supernova.

The triple-α process has reached its end with the carbon and the subsequent oxygen fusion, and at this end the carbon and oxygen fractions in stellar nucleosynthesis are about equal, a truly astounding fact.

If one looks at the triple-α process in detail, one notices how extremely interlocked the individual parameters are and how large and complex the structure of interdependencies is. Even a slight variation of only one of the many adjustment screws has effects that can hardly be compensated by a change of only a few others. The triple-α process gives the impression of an extreme fine-tuning, which is highly targeted to a life-enabling Universe. Whether this is really so and the choice of the interactions during the formation of the Universe was characterized by "wisdom", is outside the judgement of physics. One encounters here again the anthropic principle, which says in rather simple words: the Universe is as it is, if it were different, there wouldn't be any reflections about it. There is no universe to compare with. The fact that there is Life shows that the evolution of the Universe must have moved in this very direction after all.

Some reflections are nonetheless in order:

What would be,

1. if the state at 92 keV in ^{8}Be were at a higher energy?

    (a) the ^{8}Be lifetime would decrease quadratically with energy, and carbon and oxygen fusion via the triple-α process would be significantly reduced.

2. if the state at 92 keV in ^{8}Be were at a lower energy?

    (a) the ^{8}Be lifetime would increase much more than just quadratically with energy,

    (b) the triple-α process would be explosive and would rupture the star,

    (c) an explosive triple-α process could also rupture a Supergiant in its early phase and stop further nucleosynthesis to heavy elements altogether.

3. if the Hoyle state did not exist?

    (a) in fact, the Hoyle state is highly singular as far as its particular structure is concerned; it occurs in this form only in ^{12}C, and it has taken theoretical physics over 50 years to describe this state; without the Hoyle state, carbon and oxygen would only exist in tiny amounts; Life could not exist.

4. if the Hoyle state in ^{12}C were energetically a few percent either lower or higher?

    (a) being lower by 5%, the Hoyle state would have no effect at all, since it would not extend into the thermal spectrum of the helium nuclei (see Fig. 9.9); carbon and oxygen would not be formed,

    (b) being lower by 2%, the triple-α process would likely be explosive, at least much too fast; only carbon would be synthesized but almost no oxygen,

    (c) being higher by 2% or more, the influence of the Hoyle state on the synthesis of carbon and oxygen would diminish to the point where small amounts of carbon could still be produced, but likewise hardly any oxygen.

Of course, these arguments are only valid as long as the individual adjusting screws can be moved independently. This is by no means the case.

> *In a publication in 2000, different scenarios were studied both in terms of their effects on the Hoyle state and in terms of their effects on nucleosynthesis, starting from fundamental principles («first principles»). The authors show that a change in the «strong» interaction in the nucleon sector of only 0.4% is sufficient to produce either only carbon or only oxygen, but not both simultaneously[9]. Life based on carbon and oxygen would not be possible.*

9.4   Supernovae — cosmic cauldrons of the elements

The fate of most stars is pre-ordained. They end their active and largely peaceful life cycle after about $10-15$ billion years as White Dwarfs and remain in this stage almost forever — unless something unforeseen happens in their near neighborhood. Considering the time scales and the phenomenal number of new star births (in the whole Universe up to 10,000 per second, see e.g., page 48) such unforeseen events are quite probable, though.

The most spectacular events, which even affect large areas of an entire galaxy, and which are probably the most violent cosmic events after the "Big Bang", are initiated by stars that are $10-100$ times more massive than the Sun. They are at the upper end of the Hertzsprung-Russell diagram (see Fig. 9.4).

For stars of this size the life cycle does not end at all with the triple-α process. As a result of the increasing enrichment of the stellar interior with carbon (^{12}C) and oxygen (^{16}O) from the triple-α process leading to an ever more increasing gravitational compression, further fusion reactions and fusion cycles will be ignited. In these cycles, much more energy is being released per unit time, which causes the outer layers (in particular

---

[9] H. Oberhummer, A. Csótó, and H. Schlattl, *Stellar Production Rates of Carbon and Its Abundance in the Universe*, Science 289, 88 (2000).

the photosphere) of the star to once again grow significantly. The star inflates to a Supergiant, whose outer spheres can stretch to more than 1000 times the solar radius (i.e. $\simeq$ 1 billion km) over time.

The individual fusion cycles responsible for this have been uncovered under laboratory conditions in painstaking detail over the last 50 years, and the reaction rates have now all been individually measured. In particular, the Institute for Nuclear Physics in Münster distinguished itself from 1975 to 1992 by decisive and fundamental work in this field [10], which was recognized with prizes and awards.

The massive stars at the upper end of the Hertzsprung-Russell diagram run through the various evolutionary phases or fusion cycles in fast passage. This is of course also a result of their enormous energy output, which exceeds that of the Sun by a factor of $1,000-10,000$. The lifetimes of such stars are correspondingly short, typically in the range of a few ten million years, depending on their size.

The initial evolution is similar to that of normal Sun-like stars, starting with the hydrogen-cycle for helium production, then followed by helium burning according to the triple-α process. The approximate time sequences accompanied with the rising temperatures inside the star during this time are shown in Fig. 9.11.

While a Sun-like star at the end of its active period converts to a Red Giant and remains there only for a short time to finally perish unspectacularly as a White Dwarf, the end of a Supergiant has apocalyptic proportions. The final evolutionary stages of such a star proceed so fast that they are even observable on the time scale of a human life. This is detailed in the chart on the next page: The star spends about 600 years in the stage of "carbon burning", about 1 year in the stage of "neon burning", half a year in the stage of "oxygen burning", and finally only one last final day in the stage of "silicon burning". Burning shells are formed, and in each of them a specific fusion cycle takes place at a specific temperature. Heavier fusion products (called "ashes") sink into the star's interior, where they accumulate in the next shell and ignite the next cycle. With each new cycle the energy output

---

[10] Claus E. Rolfs, William S. Rodney, *Cauldrons in the Cosmos*, University of Chicago Press, 1988.

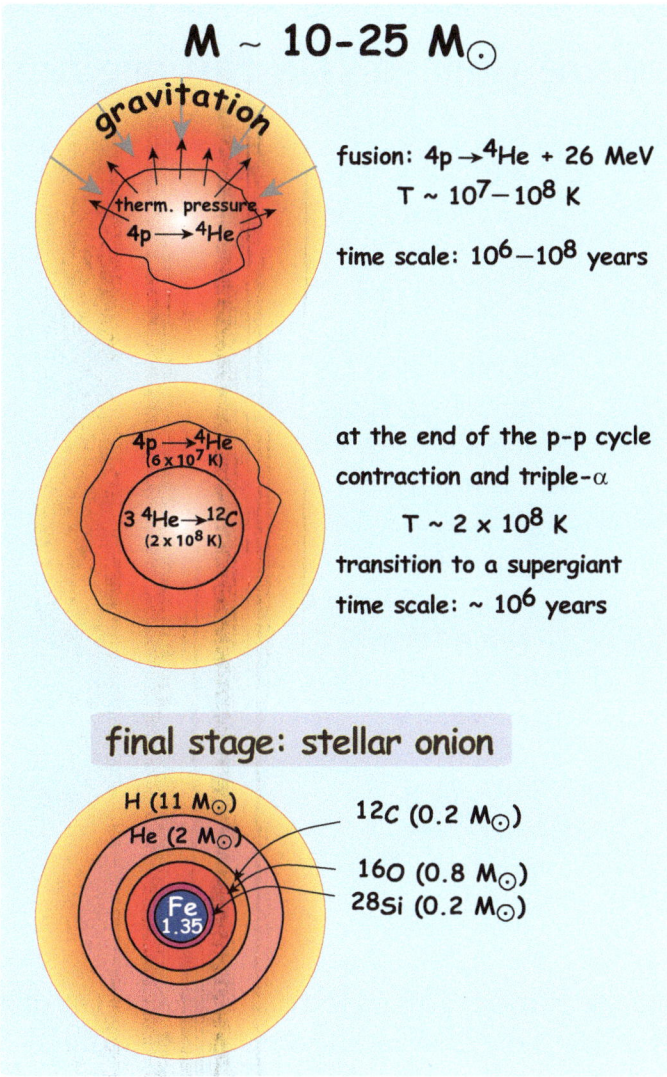

Fig. 9.11:   Stages of a Supergiant (here exemplary for masses up to about 25 solar masses). The representation is not true to scale. In the "stellar onion" the mass fractions of the individual shells are given in units of the solar mass. The iron/nickel core is about the size of the Earth, and the outer region is about the size of the solar system up to the outer planets.

per unit time increases dramatically. Lighter fusion products (e.g., hydrogen or helium) are transported by the radiation pressure back to higher layers. This continues until in the end the star has the shape of a "stellar onion", as shown in Fig. 9.11.

final phases of a Supergiant

600 years: $^{12}C + {}^{12}C \longrightarrow {}^{20}Ne + \alpha$ plus 4.62 MeV   energy

$^{12}C + {}^{12}C \longrightarrow {}^{23}Na + p$ plus 2.24 MeV   energy

... additional cycles (Ne-Na, Mg-Al) $\Longrightarrow$ energy

temperature: $(0.5-1) \times 10^9$ K

ashes: ^{24}Mg und ^{28}Si

1 year: ^{20}Ne -burning $\Longrightarrow$ energy

... additional cycles and reactions $\Longrightarrow$ energy

temperature: 2×10^9 K

ashes: mainly ^{16}O

1/2 year: ^{16}O -burning $\Longrightarrow$ energy

... additional more complex cycles

and reactions $\Longrightarrow$ energy

temperature: $(2-3) \times 10^9$ K

ashes: mainly ^{28}Si and already Fe

1 day: ^{28}Si -burning $\Longrightarrow$ energy

temperature: 4×10^9 K

ashes: mainly iron/nickel

The fusion processes reach an end with the production of the iron/nickel elements as the last stage. The nuclei in the iron/nickel region are the most stable in the periodic table, and further fusion processes are no longer possible under an energy gain. The iron/nickel elements are the "no more combustible ashes" in the stellar evolution. From now on these elements are deposited in the central interior of the star (see Fig. 9.11). Due to gravity and the still ongoing infall from higher shells, this central region gets to be more and more massive, and more and more compressed with time. At the end of this development the iron/nickel core reaches a total mass of about 1.4 solar masses.

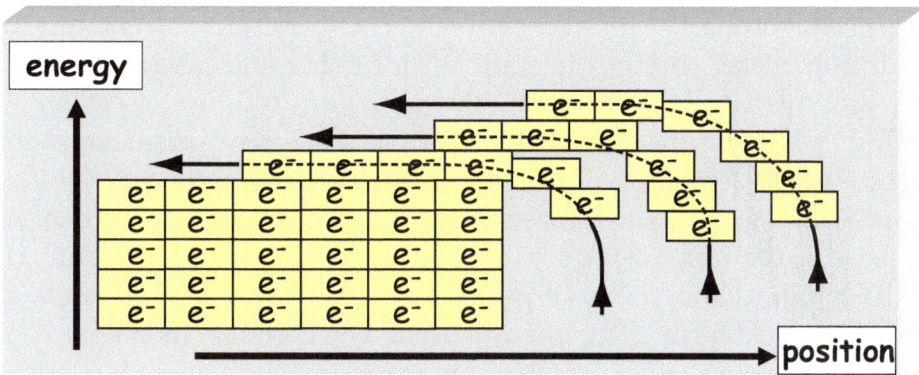

Fig. 9.12: If it becomes spatially too narrow for the electrons, some of them have to move to (energetically) higher floors. This fact still provides a (last) balance against the gravitational collapse of the inner part of a Supergiant, because in the iron/nickel core energy production from fusion has stopped and the stabilizing radiation pressure is now no longer present.

The apocalypse begins

The star is still stable. Remarkably, it is stabilized by the electrons, which have not yet appeared at all in this context. Their presence, without this having been mentioned specifically, has so far only ensured that the star, like any other stellar object, is electrically neutral to the outside. For the mass balance of a star the electrons are insignificant, they contribute less than 0.5‰ to the total mass. Their brief appearance at the present time is of a completely different quality, yet it is decisive for the sequence of events now unfolding.

Electrons are elementary particles with half-integer intrinsic angular momentum (spin), which makes them members of the "fermion" family. These possess a most remarkable property that distinguishes them from "bosons", which have integer spin. Fermions always get — somewhat casually formulated — a "single room". The size of this single room is given by the product of the spatial size Δx and the momentum size Δp, which has the value $\Delta x \cdot \Delta p \simeq \hbar$ ($\hbar = 6.582\,119\,569... \cdot 10^{-16}$ eV s). If two electrons with the same spin orientation want to occupy the same space Δx at position x, then one of them must move to a higher floor, i.e., assume a higher momentum — and thus a higher energy. This situation is explained pictorially in Fig. 9.12.

This special property initially prevents the gravitational collapse of the central iron/nickel core of the star. With further and faster mass accumulation and increased compression, electrons have to move closer and closer together, which means more and more electrons have to make room and move to "higher floors" — they become more and more energetic. The gravitational energy is thus stored temporarily in the form of electron energy. The mass of the iron/nickel core eventually reaches a critical limit, the so-called Chandrasekhar limit of about $1.3 - 1.4$ solar masses [11] at a radius of still about 6000 km. The density inside the core has increased to about 10^8 g/cm^3. The electrons in the upper levels now have enough energy to be captured by the iron/nickel atomic nuclei by means of "inverse β-decay", a process that leads to the dissociation of the nuclei and the "neutronisation" of the star in the further course. The fusion of the elements up to iron, which the star had so laboriously worked on over long periods of time, is now partially reversed. These processes cost energy, which is taken from the "upper floors" by removing their occupants.

As a result, the iron/nickel core shrinks, and in this process more and more electrons have to move up from the "lower floors" to the upper ones. The destruction process rapidly accelerates, and the entire Supergiant collapses within seconds. — A supernova is about to be born. It will outshine a whole galaxy for a few hours and will still have a brightness comparable to the galaxy for several months.

The individual stages are summarized again in the following panels:

---

[11] Named after the American astrophysicist and Nobel laureate Subrahmanyan Chandrasekhar (1910 – 1995, Nobel Prize 1983).

Supernova — collapse & explosion

1 The nuclear ashes now consisting of "non-combustible" iron/nickel accumulates inside the star. The electron pressure prevents further compression, and the radius of the iron/nickel core eventually stabilizes at about 6000 km.

2 The gravitational pressure reaches the critical limit (Chandrasekhar limit) at about 1.4 solar masses.

3 The electrons (in the upper floors) now have sufficient energy to be captured by the nuclei (inverse β-decay). The fusion elements, which the star had synthesized so far are now dissociated. This costs energy. The stability condition is thus removed and the electron pressure is released in a runaway situation within 0.5 seconds. In this process, protons and atomic nuclei literally "eat up" the electrons by means of the reaction given in this simplified version:

$$p + e^- \longrightarrow n + \nu.$$

4 Neutrinos are emitted, and at the same time the iron/nickel core becomes increasingly neutron rich (neutronization) and implodes faster and faster. In the final stage, the core consists of more than 75% neutrons.

5 The imploding matter is accelerated to 30% of the speed of light − (free fall into the center).

6 Finally, also neutrinos cannot leave the star anymore (no more cooling). The mean free path for neutrinos is now only:

$$\lambda_{mfp} \simeq 300 \text{ m} \ \ !!$$

Supernova — collapse & explosion (cont.)

7 From now on, almost all gravitational energy is converted into neutrinos. The neutrino energies are substantial and extend to about 30 MeV.

8 The collapsing matter reaches nuclear density at 30% of the speed of light.

$$\rho \simeq 10^{15} \text{g/cm}^3 \implies (1 \text{ billion tons per cm}^3).$$

The innermost volume contains about one solar mass at a radius of only about $30 - 50$ km.

9 The nuclear density acts like a hard wall; the imploding matter is thrown back at it and hardly looses energy.

10 The neutrinos are liberated and the energy temporarily stored in them is now released. A neutrino burst occurs.

11 The outgoing and extremely neutron-rich shock wave penetrates the still collapsing outer layers, and the neutrinos are heating the wave from behind!! The latter process is critical and decides whether a successful supernova explosion can indeed occur.

12 Nucleosynthesis begins when infalling and outgoing matter meet — highly neutron-rich atomic nuclei up to and beyond uranium are being produced. They eventually decay into the stable nuclei of the periodic table according to their half-lives, thereby generating the isotopic and elemental abundances shown in Figs. 9.1 and 9.2.

13 Time scale for the formation of all elements: $0.1 - 10$ seconds.

Supernova — collapse & explosion (cont.)

14 The star explodes into interstellar space with an explosion velocity of about 10,000 km/s (and up to 30,000 km/s for Supergiants, which then convert to Black Holes).
This makes available new material — now enriched in heavy elements — over large distances for subsequent formation of new stars (e.g., for the Sun 4.6 billion years ago).

Energy budget of the first $\simeq 10$ seconds

energy outflow:	$\gtrsim 3 \times 10^{46}$ joule (99% ν s)
compared with Sun:	$\gtrsim 6 \times 10^{9}$ *years of Sun*
neutrino energy:	$\gtrsim 3 \times 10^{46}$ joule
neutrino flux:	$\gtrsim 10^{57}$ neutrinos/second
compared with Sun:	$\simeq 8 \times 10^{38}$ *neutrinos/second*

Final phase

initial mass:		approx. $10-25$ solar masses
final object:	$\Longrightarrow$	neutron star, pulsar, magnetar
initial mass:		approx. $25-100$ solar masses
final object:	$\Longrightarrow$	Black Hole

9.5   Supernovae — insights & impressions

Stars that are visible to the naked eye under optimal conditions in the night sky are usually no more than about $2000 - 5000$ light-years away. This is a tiny range compared to the size of the Milky Way Galaxy. A supernova explosion, on the other hand, is visible to the naked eye over the entire range of the galaxy, and this even for weeks. Accordingly, there are historical testimonies about sudden star-appearance phenomena (so-called "guest stars"), which today can be assigned to supernova explosions. If one supposes that about 2/3 of the galactic supernova explosions remains hidden because of the opaque galactic disk, then about 3 supernova explosions per 200 years in our Milky Way can be inferred from these historical testimonies on average. Even the supernova remnants found in recent times from unwitnessed "guest stars" do little to change this statistics, so that today a reasonable assumption for a mean supernova explosion rate is 2 ± 0.8 events per century.

A few such events from the last 1000 years are worth mentioning:

For the supernova in the year 1006 (**SN1006**) there are numerous historical records from quite different parts of the world. Concerning its brightness and position in the sky it was described relatively precisely by the Egyptian physician, astrologer and astronomer Ali ibn Ridwan[12], so that NASA's Chandra satellite telescope was able to locate its present-day remains, which pointed to a former star at a distance of about 7200 light-years. As a supernova, it possessed about three times the brightness of the planet Venus and was thus easily visible during the day. A photograph taken by the Chandra telescope in the X-ray wavelength region is shown in Fig. 9.13. This supernova remnant is still expanding into interstellar space at about $10,000 \, \text{km/s}$. The absence of a neutron star or a Black Hole in the central region leads to the conclusion that SN1006 formed from a binary system of two White Dwarfs.

The **SN1054** was discovered and described by Chinese astronomers. It dates approximately to the month of July in 1054. Its remnant today

---

[12] Abu'l Hassan Ali ibn Ridwan Al-Misri (approx. $988 - 1061$) Egyptian physician, astrologer and astronomer, born in Giza.

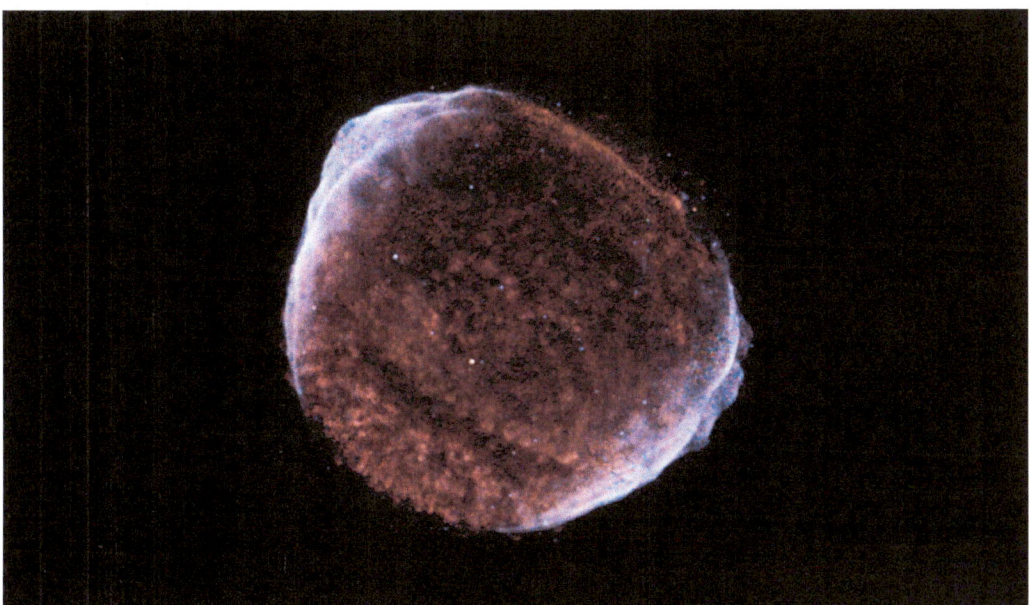

Fig. 9.13: Remnant of supernova SN1006 photographed in the X-ray range by NASA's Chandra satellite telescope. The diameter of this remnant is about 65.5 light-years, from which an expanding speed of about 10,000 km/s is calculated. [Public domain]

Fig. 9.14: Crab Nebula, the remnant of supernova SN1054. [Courtesy of NASA and Space Telescope Science Institute - STScI]

Fig. 9.15:   Remnant of the supernova SN1680. The visible diameter of the remnant is about 10 light-years (30 light-years in the X-ray spectrum). The explosion is asymmetric with jet-like expulsive velocities of more than 14,000 km/s in and against the direction of observation. At the center is an inconspicuous neutron star. [Courtesy NASA/JPL-Caltech]

forms the Crab Nebula, which also hosts the leftover neutron star. This object emits as a pulsar an extremely hard X-ray spectrum, whereby the pulsing indicates a rotational frequency of about 30 Hz. The Crab Nebula is 6500 ± 1600 light-years away and by today has a diameter of about 11 light-years. Its emitted light power is 75,000 times that of the Sun. During the explosion phase, SN1054 was probably the brightest object in the sky after the Sun (comparable to the size of the Moon) and was effortlessly visible during the day. Figure 9.14 shows an image taken by the NASA Hubble telescope.

The **SN1680** (Cassiopeia-A) (Fig. 9.15) is the remnant of a supernova explosion at a distance of 9100 light-years from the solar system. The

English astronomer John Flamsteed (1646 — 1719) may have unknowingly witnessed this event in 1680 when he prepared a catalogue of stars showing a star that later on could not be found anymore. The event is, however, interesting for a different reason. Because the explosion signal reached Earth only 341 years ago, the sequences of events like the nucleosynthesis and post-explosive processes can be studied and traced back in great detail today. Using light reflections from the interstellar gas, even the actual explosion can be visualized post-mortem again — comparable to the echo of a sound wave or the rumbling thunder after a lightning strike [13]. The supernova is classified of type IIb, i.e., the star had already lost large parts of its hydrogen envelope in the run-up to the explosion that in the end led to a reduction of the explosive power.

More supernova events from the last millennium in our galaxy are listed in the Table 9.1. All of these are still visible today as supernova remnants. In one case a late aftereffect even had a sizable impact on planet Earth, despite its distance of 47,000 light-years.

---

[13] O. Krause, et al., *The Cassiopeia A Supernova Was of Type IIb*, Science 320, 1195 (2008), see also same authors, arXiv:0805.4557v1 [astro-ph], (2008).

Tab. 9.1: Supernova events in the last millennium and some historical testimonies.

Supernova events during the last millennium		
year	distance	comment
1006	7200 ly	- witnessed in several parts of the world; - possibly brightest event ever witnessed; - best described by Ali ibn Ridwan (Egypt).
≈1100	15,000 ly	- identified in 2008 as supernova remnant; - explosion date calculated back to year ≈1100.
1054	6500 ly	- witnessed in China; - today's remnant: Crab Nebula.
1181	8500 ly	- witnessed in China & Japan in the constellation Cassiopeia — BUT: remnant until today not clearly identified.
≈1250	700 ly	- no historical testimony; - discovered in 2001 by the Chandra telescope.
1572	10,000 ly	- witnessed by Tycho Brahe → half-moon size.
1604	20,000 ly	- witnessed by Johannes Kepler → brightest star in the night sky.
≈1680	9100 ly	- named Cassiopeia-A; no historical evidence.
≈1900	27,700 ly	- discovered in 1984 as a supernova remnant; - explosion date calculated to year ≈1900.
1000–1600	47,000 ly	- explosion likely happened during this period; - remnant: strongest magnetar found so far; - on Dec. 27, 2004 this magnetar produced a 0.2 s lasting γ-ray burst that was likely caused by a stellar quake; it almost com- pletely ionized the ionosphere of the Earth and put it into oscillations; within 0.2 s the magnetar had released more energy than the Sun in 150,000 years.

Image selections

The following annotated Figures 9.16 to 9.25 give impressions about past supernova explosions and about how these explosions developed over time.

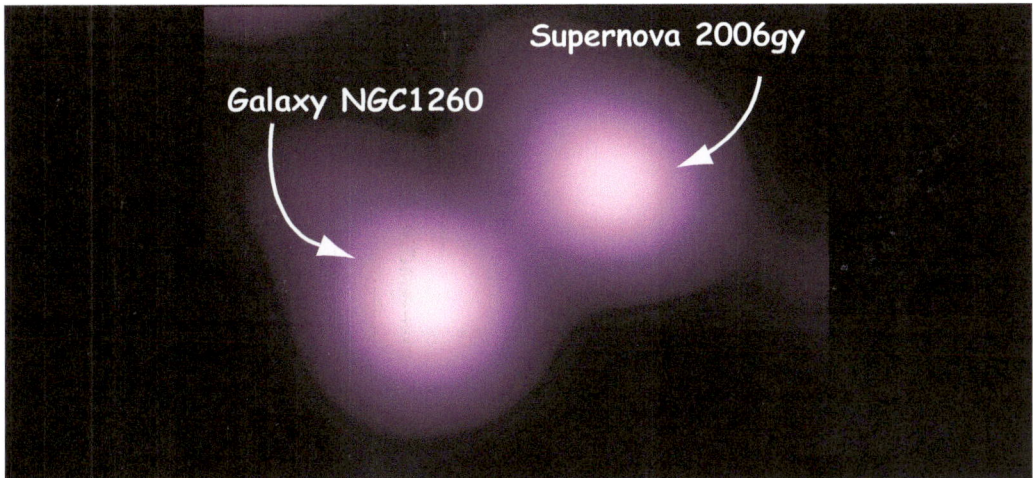

Fig. 9.16:   Supernova SN2006gy in the galaxy NGC1260, 238 million light-years away. SN2006gy is the most luminous supernova ever recorded. (Photographed by the Chandra X-ray Observatory in the X-ray spectrum) [Credit: NASA/CXC/UC Berkeley/N.Smith et al.]

Fig. 9.17:   Supernova SN2005cs in the 25 million light-years distant galaxy M51 (Whirlpool Galaxy) accidentally discovered by Wolfgang Kloehr in 2005. [CC BY-SA 2.5]

Fig. 9.18: Vela supernova remnant. The explosion occurred about 11,000 − 12,000 years ago and was only about 930 light-years away from Earth. Stone Age people will probably have suffered from sunburn for several months.

At the center is a neutron star (Vela pulsar) that pulsates with a period of 89 ms. The Vela pulsar is the closest known pulsar to Earth. The image section above covers about 20 light-years. It is assumed that the Vela supernova exploded in an interstellar nebula that was already formed by an even earlier supernova explosion. [CC BY-SA 4.0]

Fig. 9.19: Supernova remnant G299.2-2.9. The remnant is about 16,000 light-years away from Earth. The explosion occurred about 4500 years ago. The supernova belongs to type Ia, i.e., in such a case one assumes a White Dwarf extracts mass from its main companion star over time and then finally explodes as a supernova. In this case, the companion could also have been a neutron star, which survived the explosion of its partner and, by means of the shock wave, it left the event location with nearly 30,000 km/s. — The image was made from a total of 10 exposures over 5 years and is a composite image from the X-ray (Chandra telescope) and the infrared wavelength range (Two Micron All-Sky Survey). The image section covers about 114 light-years.

[Credit: X-ray: NASA/CXC/U.Texas/S.Post et al. — Infrared: 2MASS/UMass/IPAC-Caltech/NASA/NSF]

Eta-Carinae the most massive star in the Milky Way ($\sim$ 100 - 200 $M_\odot$)

50 ly

Fig. 9.20: Eta-Carinae is a double- or even triple-star system with an estimated total mass far beyond 100 solar masses ($M_\odot$). The structure with the star system inside (top image) is several light-years across and thus about 1000 times larger than the solar system. Eta-Carinae has made several attempts over the past 150 years to explode as a supernova, without success though, possibly because of the considerable mass surrounding this system (arrow in the lower image). Yet, a successful supernova explosion is foreseen in the near future. Its distance of about 7500 light-years should, however, be reasonably safe for our solar system. [Credit: Foto Eta Carinae: Jon Morse (University of Colorado) and NASA/ESA]

Fig. 9.21: **Top:** SN1987 A in the Large Magellanic Cloud before (arrow) and after the explosion. The progenitor star was a blue Supergiant (named after Nicholas Sanduleak Sk -69 202). The explosion increased its brightness by a factor of 100 million. **Bottom:** The progenitor star had ejected part of its envelope about 20,000 years before the explosion, which subsequently formed a planetary ring-nebula around it, and which is now excited to emit light by SN1987 A. [Credit: David Malin / Australian Astronomical Observatory.]

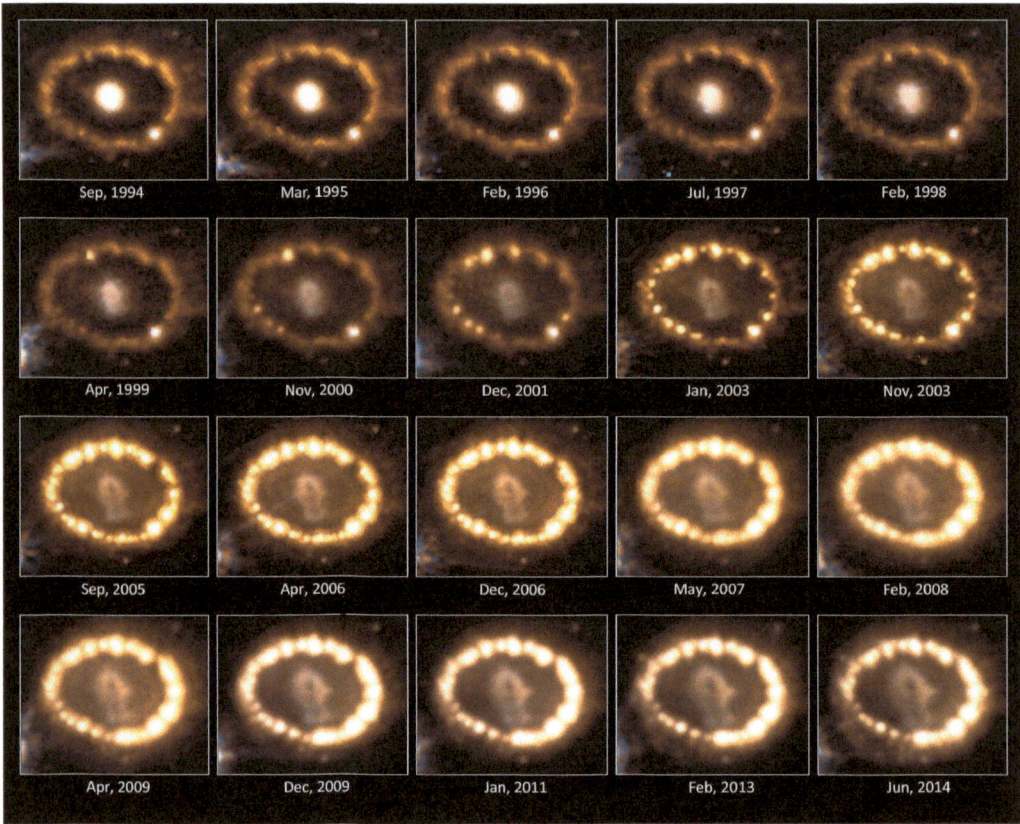

Fig. 9.22:   Ring structure around SN1987 A. Until about 2007 the brightness of the ring kept increasing and has slightly been decreasing since then. The matter shock wave coming from the supernova has not reached this ring nebula yet. This will likely happen between now and about 2030 and will lead to its total obliteration. [Credit: NASA/ESA/STScI and J. Larsson, KTH Stockholm]

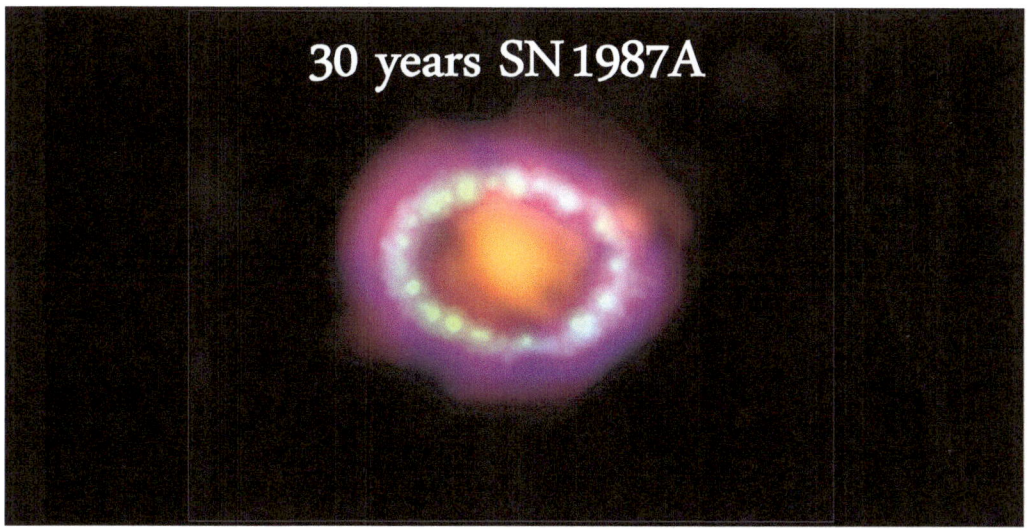

Fig. 9.23:   This image of the ring structure, which surrounds SN1987 A, was created from 4 exposures between January 2008 and January 2009. It is a composite image created from spectra in the X-ray, optical, and infrared wavelength regions. It was released to mark the 30th anniversary of SN1987A on February 24, 2017. The image covers a range of 14 light-years. [Credit: X-ray: NASA/CXC/SAO/PSU/K.Frank et al.; Optical: NASA/STScI; Millimeter: ESO/NAOJ/NRAO/ALMA]

Fig. 9.24:   The Large and Small Magellanic Clouds in the southern night sky. These two objects constitute the largest dwarf galaxies within the Milky Way's sphere of influence and the former is about 163,000 light-years and the latter about 202,000 light-years (SMC) away. The Large Magellanic Cloud hosts about 15 billion and the Small Magellanic Cloud about 5 billion stars. Together they make up about 10% of the star content of the Milky Way. To the human eye, the Magellanic Clouds look rather inconspicuous. [CC BY-SA 4.0]

Fig. 9.25: Constellation of Orion with Beteigeuze (Btg) and Rigel (Rg) in the photo.
[Credit: Constellation Orion: partially created with Stellarium under GPLv2, photo: B. Tafreshi, CC BY-SA 4.0]

The next supernova candidate?

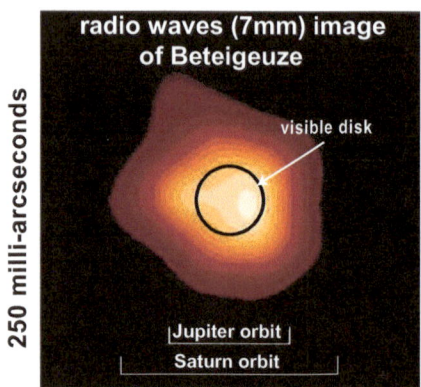

Fig. 9.26: Beteigeuze in the 7 mm radio-wave range — [Image adapted, courtesy: J. Lim, C. Carilli, S.M. White, A.J. Beasley, R.G. Marson]

With a supernova rate in our Milky Way of about one to three events per 100 years, waiting for such an event could quickly lead to fatigue. Nevertheless, there are two candidate stars that are in the final stages of their evolution. However, an exact date for their demise cannot be determined, since neither their exact masses, nor how long they have already been in this final phase are known. One star is Eta-Carinae (see Fig. 9.20), which has already launched several explosion attempts in the past 150 years, without success though; the other star is Betei(l)geuz(s)e [14] (or α-Orionis), which can easily be spotted near Orion's

---

[14] Internationally accepted spelling: "Beteigeuze"; in phonetic symbolism [ˌbetarˈɡɔrtsə]; in the English-speaking world one also hears [ˌbetelˈdʒuːs]. The name is of Arabic origin and can be translated as "giant shoulder".

belt in the night sky during the summer months (see Fig. 9.25). α-Orionis was making headlines for some time, as it apparently changed its brightness at short intervals. But this was likely not a signal for an imminent explosion; on the other hand the physics underlying this phenomenon is also not known in detail. So, ***keep your eyes open***!

The characteristic features of α-Orionis are equally impressive as those of Eta-Carinae. Its mass is estimated to be about 20 times the mass of the Sun, and its current radius is about 1000 times that of the Sun. In the solar system, its radius would extend to planet Saturn (see Fig. 9.26). Its rather close distance to the solar system of merely 640 ± 150 light-years is at the edge of getting to be worrisome. In the case of an explosion, the star's radiative power reaching Earth could grow to about 1% of the solar power for several days and nights, an effect not only easily measurable but also clearly perceptible.

9.6 Supernovae and cosmic surprises

The enormous light output of a supernova makes such a spectacular event visible to the most distant corners of the Universe. Further, the physics of the type Ia supernovae (these are supernovae which, for example, evolve from White Dwarfs and hence do not have a hydrogen shell) is by now known in detail. Their light curves can be followed for days, and they are ideally suited for determining the absolute magnitudes of such explosions. Type Ia supernovae are therefore considered "standard cosmic candles", from which cosmic distances can be deduced back to $10-11$ billion years before our time, when also star formation had been much more frequent than today.

It was for these reasons that in the late 1990s extensive observational projects were launched with the goal of detecting supernovae and thereby screening the Universe to its remotest corners in order to derive parameters such as Hubble constant, curvature, homogeneity, mass distributions, dark energy properties, to name only a few of the key properties of the Universe. These projects became known as:

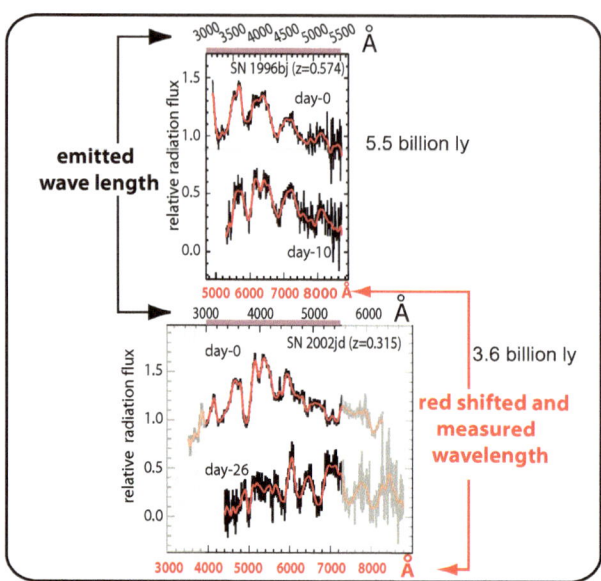

Fig. 9.27: These two examples show how spectral analyses can be used to accurately determine the redshifts of supernovae. Since the underlying physics of supernova explosions is well known, the characteristic emission spectrum can be calculated over a large wavelength range. By then applying the redshift factor $(1 + z)$, one can make the calculated spectrum (red line) coincide with the measured one. Although in the example the two supernovae are about 2 billion light-years apart, both show roughly the same emission lines. These emission lines are usually tracked for several days (for details see text).

The parts with the same wavelength range in the lower and upper images $(3000-5500 \text{ Å})$ are highlighted.   [Adapted from Blondin et al., The Astrophysical Journal 682:724 (2008)]

ESSENCE   **E**quation of **S**tate:**SupErN**ovae trace **C**osmic **E**xpansion; studied more than 200 supernovae

  SDSS   **S**loan **D**igital **S**ky **S**urvey; studied several hundred supernovae

  SNLS   **S**uper**N**ova **L**egacy **S**urvey, originated in 2003 from the SDSS program; studied more than 2000 high-redshift supernovae.

  HST   **H**ubble **S**pace **T**elescope; holds the record for the most distant supernova (at redshift $z = 2$)

Supernova redshifts and distances

It may be appropriate at this point to say a few words about how redshifts and cosmological distances are determined. The procedure sounds simple at first, but it is not trivial at all in detail. In order to determine redshifts, a line spectrum is needed, from which certain and known lines or pair of lines of atomic excitations can be identified. In a supernova explosion these are often silicon lines because silicon is produced in large quantities and appears also in the outer explosion waves. However, since an explosion wave propagates at about 10,000 km/s in the direction of the observer and, of course, in the direction opposite to the observer, atomic lines are both, blue-shifted as well as red-shifted at the same time, which leads to a significant broadening. In addition, explosion waves show different transparencies in the course of their development (i.e., over days and weeks), so that atomic transitions from deeper layers and different compositions appear with different intensities. Fortunately, these effects can be determined from a spectrum recorded over several days, so that ultimately the redshift z due to the expansion of the Universe can be safely ascertained by matching the calculated spectrum with the measured one. The procedure is exemplarily shown in Fig. 9.27 for two supernovae of very different distances. The obtained precision of better than 1% is particularly impressive.

Another equally important and by all means non-trivial conclusion can be derived from this procedure: The physics, which results from today's point of view, can be applied without recognizable modifications to times that lie more than 10 billion years in the past, or:

The laws of Nature were not subject to any temporal changes over the last 10 billion years.

Further, when calculating the distance of a far-away star based on the observed radiant power, the expansion of the Universe has to be taken into account. For example, a photon that has traveled for 10 billion years, will see a Universe doubling in size during its journey. Originally aimed at the Earth, it may now miss

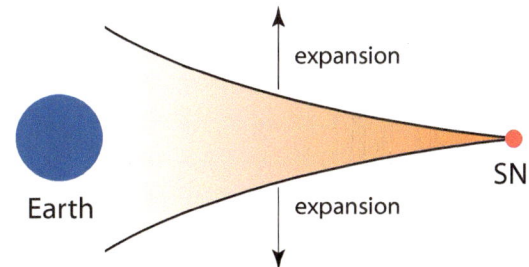

Fig. 9.28: Schematic representation of how the angle-of-view changes due to the expansion of the Universe in the case of cosmic distances (here: supernova SN – Earth).

it due to the simultaneous expansion of the angle of view. The situation is schematically explained in Fig. 9.28.

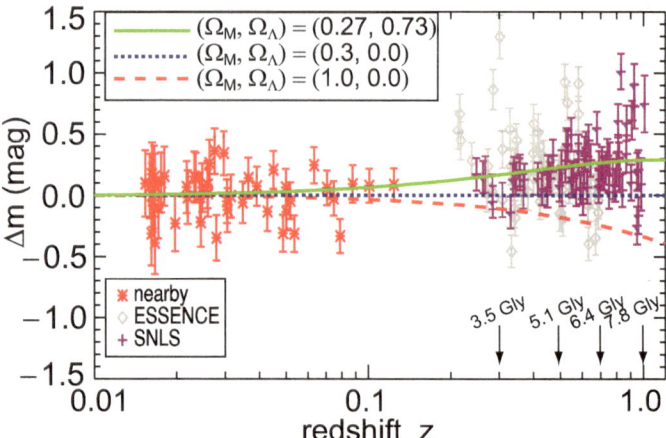

Fig. 9.29:   An amazing experimental result: From a distance of about 3 billion light-years onwards, one observes a deviation from Hubble-Lemaître law in the way that the expansion of the Universe seems to accelerate.   The light intensities show a decrease equivalent to $\Delta m \simeq 0.26$, indicating larger distances than calculated by this law[§§].   The curves show three different cosmological models; **green:** 27% matter, 73% dark energy, flat Universe, accelerating; **blue:** 30% matter, open Universe; **red:** 100% matter, flat Universe, decelerating. — **Note:** The magnitude m has a logarithmic scale, i.e., a positive Δm corresponds to a reduction in light output, in this case of about 26%. [Figure adapted from Wood-Vasey et al., The Astrophysical Journal 666:694 (2007); (CC BY-SA 4.0)]

[§§] Definition of magnitude m: If two objects differ in their flux densities l_1 and l_2 by a factor of 100, the apparent magnitudes should differ by 5, i.e., $l_1/l_2 = 100^{(m_2-m_1)/5}$ or $(m_1 - m_2) = \Delta m = -2.5 \cdot \log(l_1/l_2)$.

The systematic search for distant supernovae was particularly vigorously pursued by the American cosmologist Saul Perlmutter in the mid-1990s. His goal was to determine the so-called decelerating parameter of the Universe's expansion with the help of supernovae type Ia. The general and until then uncontested doctrine was that the expansion speed of matter, which emerged from the explosive "Big Bang" phase, should slowly decelerate as a result of gravity. However, the numerical value of this deceleration was unknown. Perlmutter succeeded in covering a much larger and deeper area of the sky through a series of technical improvements to existing telescopes, thereby increasing many times over the number of supernovae observed during an observation period. By year 2000, this development had earned him a significant volume of data from supernovae up to redshifts $z = 0.5$, a world distance-record at the time.

The bafflement was immense. The expected deceleration turned out to be an acceleration. At the same time, Brian Schmidt and Adam Riess also came to the same conclusion based on their own observations, for which all three were awarded the Nobel Prize in Physics in 2011 [15].

Today's Universe is obviously expanding faster and faster — the only question is "who is pulling on the other side?" The term "dark energy" became legitimate from then on. First calculations with a cosmological constant (dark energy) with an anti-gravitational effect seemed to reproduce the data. This is already recognizable in Fig. 9.29 by taking the early data sets accumulated up to 2005. In the figure, three different cosmological models are compared:

1. for a Universe that has a critical mass ($\Omega_M = 1$), the observable expansion slows down as expected and finally comes to a standstill with time $t \to \infty$ (red dashed curve);

2. for an open Universe (here: $\Omega_M = 0.3$), the observable expansion continues to grow with ultimately constant velocity (blue dotted curve);

3. for a Universe with $\Omega_M = 0.27$ and an anti-gravitational (dark energy) fraction of $\Omega_\Lambda = 0.73$, the observable expansion accelerates; this is the situation that reproduces the experimental data (green solid curve).

In this early data analysis, the values $\Omega_M = 0.27$ and $\Omega_\Lambda = 0.73$ were taken from the first results of the Planck experiment (measurement of the cosmic background radiation, see chapter 8) and were not determined independently from the supernova observations. With the significant increase of the data volume in the following 10 years, an independent analysis finally became possible. The data situation in 2014 is shown in Fig. 9.30. A total of 740 supernovae up to redshifts of $z = 2$ ($\sim$ 10.5 billion light-years) are recorded here and subjected to the analysis. The significance of an accelerating expanding Universe is now convincing and unmistakable. The matter content of the Universe determined independently from these new

---

[15] 2011 Nobel Prize in Physics: Saul Perlmutter, Lawrence Berkeley National Laboratory, Berkeley; Brian Schmidt, Australian National University, Canberra; Adam Riess, Johns Hopkins University, Baltimore.

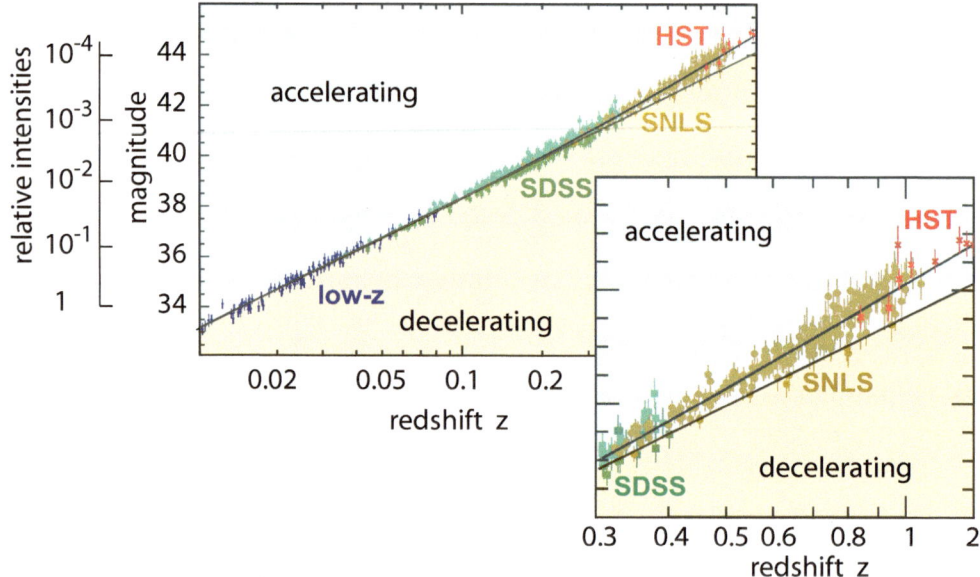

Fig. 9.30:   This figure summarizes 740 supernovae of type Ia from different observation series. With the statistical significance now considerably improved, the effect of an accelerated Universe is clearly highlighted. [Figure adapted from the data of the projects mentioned; low-z data from: Mario Hamuy et al., The Astronomical Journal 109, 1 (1995) and P. M. Garnavich et al., The Astrophysical Journal, 493:L53 (1998)]

data gives a value of $\Omega_M = 29.5 \pm 3.4\%$ and consequently for the dark energy a value of $\Omega_\Lambda = 70.5 \pm 3.4\%$. The values published by Planck and WMAP in 2019 are: $\Omega_M = 31.1 \pm 0.6\%$ and $\Omega_\Lambda = 68.9 \pm 0.6\%$ — an amazing agreement.

Considering now that, in the one case, these values are encoded in the microwave background radiation, which originated 380,000 years after the "Big Bang"; in the other one, they are encoded in the accelerated expansion of the Universe, which took place over a period of more than 10 billion years. One must state again with some amazement:

Nature did not negotiate its laws anymore at any time.

To prevent misunderstandings: Today we see objects at large distances and in the far past moving away from us in an accelerated way. Conversely, this means that acceleration has seized us today; it is the Universe of today that is expanding at an increasing rate in relation to these distant objects.

So what is "dark energy"?— Unfortunately, there is no comprehensive answer to this question yet. Speculations are therefore allowed. However, there are already some undisputed properties that characterize dark energy:

1. Dark energy is not formed from mass particles; the speculation that it is antimatter, which has an anti-gravitational effect, can safely be rejected.

2. Dark energy has the property of a cosmological constant — this follows from both the supernova data and the Planck data; its energy density is equivalent to a "negative pressure", i.e., a tension.

3. Dark energy appears to be homogeneously distributed throughout the Universe; local inhomogeneities due to interactions with normal matter (e.g., galaxies) are not observed.

4. To assume dark energy is a remnant from the inflationary phase is plausible — but a theoretical construct for such an assumption does not exist so far.

5. Dark energy contributes to the total energy density of the Universe in such a way that the sum of the mass-energy density (calculated from the mass equivalent $E = Mc^2$) and the dark-energy density equals the critical density $\Omega_{\mathrm{crit}} = 1$; the Universe is thus flat or Euclidean. This property already followed from the Planck/WMAP data on microwave background radiation and is now also derived from the supernova data presented here.

9.7 Supernovae and the creation of Life

Supernovae have apocalyptic effects on large and extended galactic regions that may span several hundred light-years. Any life that would be existent in these regions is wiped out. On the other hand, supernovae also produce the elements that are in every way indispensable for a development of Life and its variety of forms. This applies in particular to the solar system, which formed from the remnants of a supernova explosion about 4.6 billion years ago. However the star, which suffered this explosive fate and which possibly perished as a neutron star, has not been identified until today. It is conceivable that it was catapulted out of the galactic plane.

That supernovae are also directly responsible for the origin of Life sounds a bit far-fetched. We will nevertheless pursue this assumption and use some of the reasonings as they first appeared in a publication in 2010 [16].

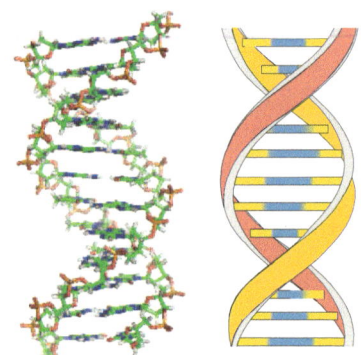

We start with a set of simple facts, by which we try to characterize the notion of «*Life*» from a purely physics point of view:

Fig. 9.31: The DNA double helix is synonymous with «*Life*». Hydrogen bonds at the protein-ends connect the two single strands.

▶ «*Life*» is an unstable state outside the thermodynamic equilibrium. To maintain such a state, which is characterized by a lower entropy or a higher order, the laws of thermodynamics demand a constant and continuous energy supply. A fraction of this energy must always be released back into the heat bath of the environment. When this non-equilibrium state is no longer maintained, Life ceases. The state, resp. the object, falls back into the thermodynamic equilibrium and takes on the temperature of the environment. The same is true as soon as the energy supply ends. Only a timely reproduction ensures continuance.

▶ Evolution is possible only if the environmental conditions keep changing. On Earth this was ensured already during the early phases of evolution, i.e., by the day/night change, by the changing climate, but also and especially by the tides of the oceans, which are propelled by the changing position of Earth's moon.

---

[16] R. N. Boyd, T. Kajino, and T. Onaka, *Supernovae and the Chirality of the Amino-Acids*, Astrobiology 10, 561 (2010).

▶   A living organism is biochemically made of proteinogenic (protein-forming) amino acids, which arrange themselves as proteins within the DNA in the form of a right-handed helix. This is also where replication takes place (see e.g., Fig. 9.31).

▶   The number of proteinogenic amino acids needed for Life is 21. All forms of Life, whether plant or animal, bacterium or virus, possess only these 21 amino acids, which they use as the building blocks for the proteins in the DNA (or RNA). The number of theoretically possible amino acids is, however, unlimited. The term "handedness" is defined by the position of the nitrogen atom in the chemical compound. (Chemists know this as Fischer projection.) Amino acids are also optically active, i.e., they rotate the polarization of the incident light (right, i.e., clockwise or left, i.e., counterclockwise) — however, this property will not be discussed here in further detail.

▶   Amino acids (with the exception of the simplest one, glycine, which is non-chiral) possess a "handedness", or in scientific terms, a chirality; they therefore belong to the group of enantiomers. They can be compared to right and left gloves, but otherwise they are completely identical. Chemically, however, they are not interchangeable, just as a right glove cannot be interchanged with a left one.

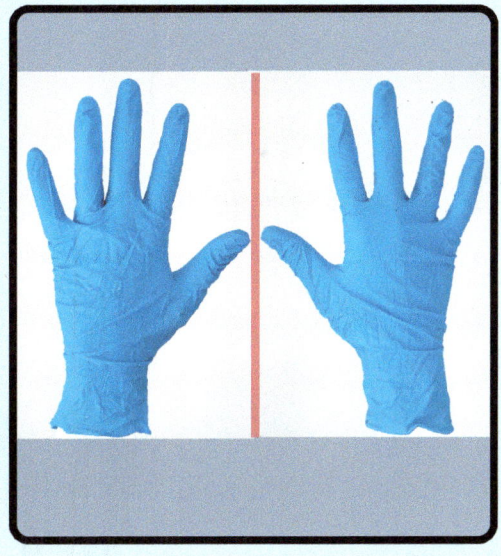

▶ Another ordering parameter of Life is added at this point: living organisms possess without exception left-handed proteinogenic amino acids. Even more astounding: evolution, in spite of its diversity, has adopted this asymmetry at all stages from the very beginning and has not changed it ever. An organism that does not possess or loses this asymmetry, is not viable and cannot develop into a living form.

▶ Right-handed proteinogenic amino acids cannot be incorporated into living organisms — they do not dock onto existing chemical compounds, like a right glove does not fit on a left hand. Right-handed amino acids are recognizable to humans: they don't taste, smell bad and are sometimes even harmful.

alanine - the simplest amino acid

L-alanine   mirror   D-alanine

▶ With the death of a living organism, racemization sets in. This process describes the conversion of an L enantiomer (for "laevus" ≜ optically levorotatory) into the corresponding D enantiomer (for "dexter" ≜ optically dextrorotatory) until the resulting racemic mixture consists of equal proportions of both left- and right-handed variants. The time constants for this transition are not uniform, they depend on many different factors. They can range from minutes to billions of years. With the transition, the ordering parameter decreases and the entropy as a measure of disorder increases again. Organisms lose their ability to reproduce. Seeds, for example, do not have an unlimited germination capacity.

▶   The state of higher order (or lower entropy) cannot arise by itself out of a thermodynamic equilibrium. An external inducement is mandatory. This also means that the early environmental conditions in the history of Earth cannot produce this asymmetry by itself. However, before this sounds too dogmatic: the «*weak*» interaction creates a tiny window of exception.

A brief excursion: Enantiomers occur in many complex organic compounds. Although they are mirror images of each other, their chemical properties are sometimes very different. This often poses a high risk for pharmaceutical companies. A particularly extreme case may be recalled from about 60 years ago, when the pharmaceutical thalidomide was launched on the market under the name Contergan, after it had gone through many safety checks to obtain market clearance. Thalidomide is an enantiomer, whose right-handed variant has a sedative effect and is harmless, and whose left-handed variant causes malformations in human embryos, and only in these. It came to market as a racemic mixture, but even an enantiopure application would not have prevented its effect on human embryos, since racemization occurs in the human body within a few hours. Note the figure.

Back to the left-handedness of amino acids. The described peculiarities concerning their chiral occurrence raise a whole series of questions:

▶   Did left-handedness arise by chance?

▶   Is Life conceivable on the basis of right-handed amino acids?

▶   By what mechanism was Earth "inoculated" with left-handedness? And why was this mechanism so extremely selective? — Note: An "inoculation" with both a right- and a left-handedness at different places and/or at different times would be fatal for the further development. All life forms would arise in duplicate, but would have to be kept away from each other at all times. There would be two different biological worlds, which would be maximally incompatible with each other.

▶   Is the emergence of Life on Earth of cosmic origin? — and if so, what is the mechanism?

Evidence that the building blocks of Life on Earth are of extraterrestrial origin has been emerging since the mid-1990s, and is by now being question-ed less and less. A new branch of science, "astrobiology", has emerged as a result. Astrobiology addresses the question of whether the cosmos may even be an effective chemistry laboratory, in which the carbon-based macromole-cules necessary for the initiation of Life are synthesized. Numerous and highly diverse organic compounds found in solar and extrasolar meteoric rocks and meteoric dusts, trapped and preserved there over long periods of time, provide overwhelming evidence that complex chemical reactions do indeed occur. They operate, of course, over cosmic time scales and strangely enough even at temperatures near absolute zero. To initiate such chemical reactions at these temperatures, clearly a different type of an "energy-provider" must be present, and only cosmic rays, both ionizing and non-ionizing, come here to mind. These flood the entire galaxy, and the radiant objects, such as stars, neutron stars, or even supernovae indeed provide a constant and never-ending supply of energy at all levels. This rather exotic interplay of chemistry and cosmic radiation can now be simulated in a laboratory under comparable cosmic conditions and can be studied in a targeted manner, naturally by shortening cosmic time scales.

Meanwhile, the branch of astrobiology science is well established, and even offered as an advanced study program at universities.

But the central question still remains:

What mechanism breaks the right-left symmetry and is thus responsible for the left-handedness of proteinogenic amino acids?

This must be necessarily a process, which itself is not right-left symmetric, respectively not mirror-invariant.

To a particle physicist the «*weak*» interaction immediately comes to mind. It violates mirror invariance, but unfortunately only marginally for the present cases. The hope having found a solution here therefore quickly dissipates — the «*weak*» interaction is simply too weak by orders of magnitude. Calculations show [17] that the effect of symmetry breaking is at best in the order 10^{-18}, i.e., among about 10^{18} molecules there is on average just one without a chiral partner as a consequence of the interaction. However, the authors of reference [17] find that indeed some of the L-amino acids are favored over their D-variants, although it is not clear whether this applies to all proteinogenic amino acids. The authors also point out that, under certain chemical conditions, there could be an effective (and also well known) amplification mechanism once an initial imbalance has been established. Whether this is still valid for an imbalance of the order of 10^{-18}, and given that in a realistic environment additional non-specifiable foreign molecules are also present in significantly higher concentrations, has not yet experimentally been proven and is somewhat doubted. But at least this is a process that would not have to be extraterrestrial.

In a little-noticed paper of J. van Klinken of the University of Groningen [18] an extraterrestrial process is discussed, which brings an additional variant into play that was not mentioned in an earlier publication by Rubenstein et al. [19] According to the scenario of van Klinken, at the time when the solar system with its planets slowly (i.e., over a period of some 100 million years) forms from the remnants of a supernova about 4.6 billion years ago, another supernova explosion event occurs nearby. A leftover rotating neutron star

---

[17] G. E. Tranter, *The parity-violating energy difference between enantiomeric reactions*, Chemical Physics Letters 115, 286 (1985). See as well: G. E. Tranter, A. G. Macdermott, *Parity-violating energy differences between chiral conformations of tetrahydrofuran, a model system for sugars*, Chemical Physics Letters 130, 120 (1986).

[18] J. van Klinken, *Broken symmetries at the origin of matter, at the origin of life and at the origin of culture.* Acta Physical Polonica B29, 11 (1998).

[19] E. Rubenstein, et al., *Supernovae and Life*, Nature 306, 118 (1983).

now emits intense and polarized UV radiation in the direction of the forming solar system. It acts (again over cosmic time scales) on the gradually evolving chemistry there, and many different organic compounds find their way into the oceans now forming on Earth. Laboratory studies show that the polarized UV radiation simply destroys enantiomeric compounds to a large part. However, it does this in a somewhat selective way, so that the remaining fraction indeed shows an enantiomeric asymmetry. Thus, the basic building blocks of Life could have been fabricated, from which over the next billion years the simplest life forms on Earth could in principle have developed (i.e., about 3.6 billion years before our time). If one assumes such a scenario, then the amino acids in the entire solar system (if they are not racemized) should occur with a uniform handedness. That would be an important indication of the validity of this hypothesis. Unfortunately, so far no amino acids have been found, not even on Mars. Further, in this picture the left-handedness is accidental because the polarization of the radiation depends on the rotation direction and the direction of the magnetic field of the neutron star, and these are not pre-determined.

The hypothesis of a second supernova would indeed be testable by ultra-sensitive mass spectroscopy looking for radioactive elements in indigenous moon rocks, which were implanted into the rocks as a result of a near-by supernova and which have not fully decayed yet. The best candidate is the trans-uranium element plutonium-244 with a half-life of 80.8 million years. Its survival fraction after 4.6 billion years is $\approx 10^{-17}$. However, since the hypothesis of a second supernova seems a bit too speculative, we will not go into these details here.

It was already said that meteoric rock and meteoric dust contain important messages about past and present cosmic processes. The majority of these messengers are, however, part of our solar system and were probably sent on a journey towards Earth through collisions in its outer areas. The increase in material on the Earth as a result of these processes is impressive, and according to an estimate by NASA could be several thousand tons per day. Most of these objects, though, do not reach the Earth's surface and already burn up in the upper atmosphere. It may be worth mentioning that in order to support research in this direction, many commercial airliners often have meteor-dust collectors under their wings, so there should be no shortage of meteoric material. Extrasolar or even interstellar meteorites are comparatively rare, and extragalactic objects, those coming from outside our Milky Way Galaxy, are highly unlikely to be found.

A meteorite, which has been making headlines for some time, is the Murchison meteorite. It descended in Australia in 1969. This 100 kg chunk disintegrated upon entering the atmosphere, and many of the fragments have been collected and since been sent to various research laboratories and museums (see Fig. 9.32). The age of the meteorite was determined to be 4.96 billion years. It is therefore presolar and thus an important witness of the events before the formation of the solar system.

Fig. 9.32:   Fragment from the Murchison meteorite shower in the National Museum of Natural History (Washington) [CC BY-SA 3.0]

On its long journey it has also collected cosmic material ("stardust"), some of it up to $7-8$ billion years old, i.e., almost twice the age of the solar system [20]. This material is the oldest solid material found on Earth so far (see Fig. 9.33).

The meteorite owes its fame to a completely different feature: Because of its carbon richness, it has been an extremely effective cosmic chemistry laboratory. It contains a large number of

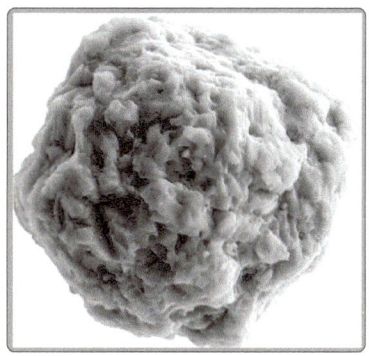

Fig. 9.33: Presolar silicon carbide grain dissected from the Murchison meteorite. The size of the grain is about 9.6 μm.
[image credit: P. R. Heck et al., Proc. Natl. Acad. Sci. USA, 1884 (2020), CC BY-NC-ND 4.0]

---

[20] Philipp R. Heck, et al., *Lifetimes of interstellar dust from cosmic ray exposure – ages of presolar silicon carbide*, Proceedings of the National Academy of Science of the USA (PNAS) 117(4), 1884 (2020).

organic compounds, including some that do not exist in their natural form on Earth. Amino acids were also produced in abundance during its long years in space: Over 80 different amino acids have been identified in the meteorite so far (and the trend is still rising), including significant quantities of those that are considered to be the building blocks of terrestrial life. The majority of them are, however, abiotic and do not occur on Earth, which was ultimately also a reason why the meteorite spread a foul and unknown odor when it fell to Earth.

The remarkable thing is that the amino acids, if not racemized, are found to be predominantly left-handed, with an enantiomeric excess of sometimes 15%. Apparently, the building blocks of Life seem to have been present long before the formation of the solar system. Maybe left-handedness was not a coincidence after all, only how and where did it originate???

As an example, an abiotic amino acid found in the Murchison meteorite is methylisoleucine. It is similar in chemical structure to the proteinogenic amino acid isoleucine. Due to the additional methyl group CH_3, it has now two chiral centers and thus four stereo-isomers as shown in Fig. 9.34 (which should be correctly named as the L and D variants of α-methylisoleucine and α-methylalloisoleucine). In the Murchison meteorite, there is a clear excess of the LL and LR variants, each in the range of 10%. This also means that in this case the time constant of racemization was in the order of the age of the meteorite. Since the LL variant racemizes primarily to LR, the formation of this amino acid must have obviously favored the LL variant over all others. Such a scenario cannot be explained by the action of polarized UV radiation.

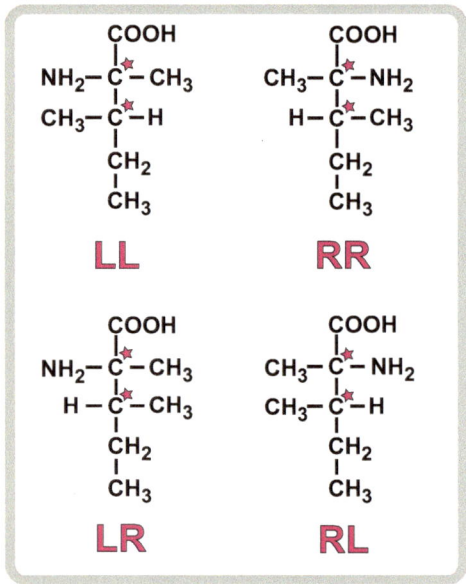

Fig. 9.34: Methylisoleucine with 2 chiral centers (indicated by the asterisks) and the 4 stereo-isomers (characterized here as "left"(L) and "right"(R)). In Fischer projection shown here, the complicated 3-dim structure is projected to 2 dimensions according to a fixed prescription.

In the course of investigating the cause, the authors of reference [21] present a startling hypothesis:

Supernovae and the resulting neutron stars are causing the left-handedness of amino acids in a direct and universal manner — randomness is excluded.

The idea behind this is that supernovae and the emerging neutron stars are one of the most intense cosmic neutrino sources, with an intensity at least 10^{15} times that of the Sun. (As a reminder, the number of solar neutrinos arriving on Earth per cm^2 and second is about 10^{10}). Neutrinos are also particles that have an intrinsic angular momentum of $J = 1/2$ and thus, they possess a handedness — left-handed neutrinos have a left-handed intrinsic angular momentum and are called "neutrinos", the right-handed ones have a right-handed intrinsic angular momentum and are called "anti-neutrinos". The left- and right-handedness is related to the direction of flight. Both forms interact with matter in totally different ways.

The chain of reasoning is fascinating and brings together a number of completely different, though experimentally largely validated effects. To follow this line of reasoning further, let us first look at the interaction with nitrogen, the central and defining atom of all amino acids:

(1) nitrogen + neutrino $\quad {}^{14}N + \nu \quad \rightarrow \quad {}^{14}O + e^- \quad E_{min} = 5.15\,MeV$
(2) nitrogen + anti-neutrino $\quad {}^{14}N + \bar{\nu} \quad \rightarrow \quad {}^{14}C + e^+ \quad E_{min} = 1.18\,MeV$

Both reactions destroy the molecule: In the first case (1) the neutrinos transform the nitrogen nucleus into an oxygen nucleus and in the second case (2) the anti-neutrinos transform the nitrogen nucleus into a carbon nucleus. (To note: Nature disallows an interchange of these reactions!) The minimum energies required for these reactions are different; for the first reaction (1) the minimum energy is $5.15\,MeV$ and for the reaction (2) it is $1.18\,MeV$. Since supernova neutrinos have energies of about $10\,MeV$ on average, reaction (1) is suppressed by about a factor of 20 compared to reaction (2). This strong suppression is essential for the further argument, i.e., in the following it is sufficient to look only at the reaction with the anti-neutrino. The rationale becomes more transparent by this.

---

[21] R. N. Boyd, T. Kajino und T. Onaka, *Supernovae and the Chirality of the Amino-Acids*, Astrobiology 10, 561 (2010), see as well same authors, *Supernovae, Neutrinos and the Chirality of the Amino Acids*, Int. Journal of Molecular Science 12, 3432 (2011).

In the presence of a magnetic field (which is generated, for example, by a neutron star as a consequence of a supernova) a special feature comes into play. The atomic nucleus nitrogen-14 is one of only three stable atomic nuclei with even mass number A and an intrinsic angular momentum (the other two are deuterium (A=2) and lithium-6 (A=6)). They all carry the angular momentum $J = 1$. Associated with the angular momentum is a magnetic moment. This magnetic moment aligns in the magnetic field along the magnetic field lines. The reaction probability with anti-neutrinos (we leave out the neutrinos as said above) now depends on the orientation of the angular momentum of the nitrogen-14 nucleus with respect to the direction of motion of the anti-neutrino. This situation is explained in Fig. 9.35. If the orientation of the nitrogen nucleus and the direction of motion of the anti-neutrinos are the same (as at the exit of the magnetic field in Fig. 9.35), then the above reaction (2) is suppressed by an order of magnitude compared to the reverse case (as at the entrance of the magnetic field in Fig. 9.35).

However, the handedness of a molecule actually has little to do with the orientation of the atomic nucleus. The handedness is exclusively a property of the spatial structure of the electron cloud in the molecule. In this respect, the arguments in the previous paragraph could apply equally to both handednesses, and the net effect would be zero.

Nonetheless, there is a communication between electrons in the molecule and the angular momentum (spin) of an atomic nucleus, albeit weak. It goes under the name of hyperfine interaction. While this is not quite applicable here in the classical sense, the authors in reference [22] show that the electron cloud of the entire molecule does indeed align itself in a magnetic field along with the nuclear spin! This creates a situation as shown in the lower part of Fig. 9.35. If anti-neutrino spin and nuclear spin are aligned (left image), the reaction rate with anti-neutrinos is reduced compared to an anti-aligned situation (right image), which means on the left image side the R-enantiomer is mostly destroyed, and on the right image side the L-enantiomer is mostly destroyed.

---

[22] A. D. Buckingham und P. Fischer, *Direct chiral discrimination in NMR spectroscopy*, Chemical Physics 324, 111 (2006) and
A. D. Buckingham, *Chirality in NMR spectroscopy*, Chemical Physics Letters 398, 1 (2004).

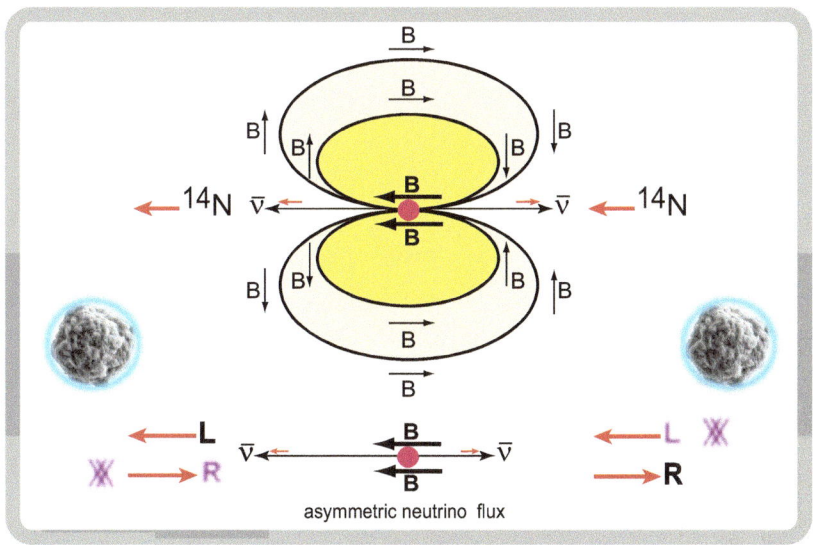

Fig. 9.35: Neutron star (red dot) with its magnetic field.
Upper part: Polarization of ^{14}N in the magnetic field. The spin directions are indicated by red arrows. The neutron star born from a supernova is an intense neutrino emitter, with neutrinos emitted primarily along the magnetic field axis. For anti-neutrinos the spin direction (red arrows) is always parallel to their direction of motion, for neutrinos (not shown here) it is always antiparallel. If anti-neutrino spin and nuclear spin are aligned (left side), the reaction rate with anti-neutrinos is reduced compared to the situation on the right side of the picture.
Bottom part: ^{14}N polarization in L- and R-amino acids as a result of hyperfine interaction. Reaction with anti-neutrinos, and thus destruction of the molecule, occurs primarily for the R-enantiomer on the left side of the image and primarily for the L-enantiomer on the right side of the image. Due to the asymmetric neutrino emission of a neutron star, there is an excess of L-enantiomers in the net balance, which subsequently mix with interstellar material.

Crucial for the formation of a chiral asymmetry is an asymmetry of the neutrino emission. Indeed, in the course of the formation of a neutron star, neutrinos (of both types) are emitted mainly in the direction of the magnetic field. This is a consequence of a higher neutrino transparency for this emission direction [23]. The asymmetry is about $1-10\%$, and since about $10^{53} - 10^{54}$ neutrinos are emitted, the neutron star gets even an additional "kick", which accelerates it to about $300-500\,\mathrm{km/s}$, enough to leave a galaxy — in some 10 million years.

In the case of asymmetric neutrino emission, the process of converting

[23] C. J. Horowitz und Gang Li, *Cumulative Parity Violation in Supernovae*, Physical Review Letters 80, 3694 (1998).

nitrogen to carbon (see reaction (2) on page 173) is reduced in Fig. 9.35 (right), and the L-enantiomer is now preferred in the overall balance. Therefore, a chiral asymmetry will be imprinted in interstellar objects (e.g., nebulae, dust, or meteorites), which have already synthesized (or collected) organic compounds and eventually find themselves within the range of the neutron star's magnetic field. The magnitude of this asymmetry is roughly estimated to be about $10^{-7} - 10^{-9}$ per neutron star encounter. Since the process is cumulative, a presolar meteorite like the Murchison meteorite, which might have passed a few 100,000 neutron stars over time with each of them leaving a fingerprint, could have accumulated a considerable asymmetry in its organic compounds.

In this particular scenario, the origin of the left-handed proteinogenic amino acids is universal. But this also means:

> **If the left-handedness has already been established at an early time of the cosmic evolution and if Earth has been inoculated with left-handedness through a cosmic insertion, then such an insertion must also have occurred on planets outside the solar system and, under Earth-like conditions, must have produced Life.**

Quite complicated but by no means excluded — <u>AND:</u> Neutrinos, which actually would be totally useless cosmically seen, could finally get some meaningful function.

Once again in a more comprehensible short form. Three different scenarios were presented that could conceivably be the origin of the chiral asymmetry of proteinogenic amino acids:

▶ **The «*weak*» interaction:** It is a universal, mirror-invariance-breaking interaction. It creates a chiral asymmetry and gives one enantiomer a tiny thermodynamically induced advantage over the other. To the degree to which this advantage must always lie with the L-enantiomer in the case of proteinogenic amino acids has not yet been demonstrated, nor is it at all obvious. Among the three different scenarios, this is the only one that is not extra-terrestrial.

▶ **Polarized UV radiation:** As an extra-solar source, it acts during or shortly after the formation of the solar system on organic molecules in proto-planetary nebulae and produces a chiral asymmetry. The formation of the L-enantiomers on Earth (and throughout the solar system) was then by chance. The fact that pre-solar meteorites like the Murchison meteorite and pre-solar meteorite chunks found in Antarctica also show a surplus of L-amino acids makes this coincidence rather improbable.

▶ **Reactions with neutrinos:** After their formation from supernovae, neutron stars are for a long time, (i.e., in the order 100,000 years) the most intense cosmic neutrino sources known. The extreme magnetic fields (up to 10^{20} times the Earth's magnetic field) of rotating neutron stars cause a focusing of the neutrino emission along their rotation axis and at the same time an alignment of amino acids, which are already present in compact objects like dust or meteorites. The alignment occurs by means of a "hyperfine-like" coupling of the nuclear magnetic moment of the ^{14}N nucleus with the electron cloud of the molecule. Finally, the decisive factor for a left-handed amino acid excess is the asymmetric neutrino emission from the neutron star along the magnetic field axis. This mechanism is in fact universal, and by means of a constant "meteorite inoculation" over cosmic time scales, the evolution of Life based on left-handed proteinogenic amino acids seems natural and likely on all Earth-like planets, solar or extra-solar alike.

The extent to which one or more, or maybe even none of these scenarios is ultimately decisive for the production of the building blocks of Life cannot be answered at present. They all have in common that the initial enantiomeric asymmetry is small and that an effective and self-replicating amplification mechanism is required. Such a mechanism, however, is strongly dependent on the environment into which an enantiomer is inserted. In particular, water content as well as surface properties of meteoritic rocks play a decisive role.

A quote from William Bonner [24]:

"All conceived theories to amplify molecular chirality have racemization as their ultimate Nemesis."

---

[24] W. A. Bonner, *The origin and amplification of biomolecular chirality*, Origin of Life and Evolution of the Biosphere 21, 59 (1991).

9.8   In search of Life in the solar system

"In search of Life in the solar system" — this is the motto emblazed upon almost every interplanetary mission. The chances of directly encountering "living" Life on other celestial bodies in the solar system are miniscule; far more successful though, might be the search for traces of past life forms. But even the discovery of left-handed amino acids on another planet would be an event that would decisively change the existing worldview.

For over 60 years, the interplanetary target in this search has been our neighboring planet, Mars (see Fig. 9.36). Mars has properties that are similar to Earth in many respects. It could well be capable of developing and sustaining primitive forms of life. Mars, however, is only about half the size of Earth (Mars: $\emptyset \simeq 6,800$ km, Earth: $\emptyset \simeq 12,700$ km), and its gravitational pull is only about 38% of that of Earth. This is just enough for it to bind the thin atmosphere of currently about 6 mbar (compared to Earth: 1013 mbar). Unfortunately, the low density of the atmosphere also means that an efficient temperature equalization between day and night, as it occurs on Earth due to weather, is not possible. Temperature fluctuations from $-85°$ C to $+20°$ C are the consequence. More favorable in this context is Mars's rather fast self-rotation of just under 24 hours, which prevents even larger fluctuations.

Mars's thin atmosphere consists mostly of carbon dioxide (96%) with traces of nitrogen (1.9%), argon (1.9%), oxygen (0.15%), and water (0.02%). Large reservoirs of water are found as ice at the poles, enough to fill large Martian oceans. This means that the basic building blocks for the formation of organisms on Mars are abundant, and environmental conditions can be considered reasonably favorable.

The fact that Mars still has a thin atmosphere suggests that the atmosphere may have been much denser shortly after the formation of Mars about 4.5 billion years ago. Due to the low gravitational pull and the lack of a protective magnetic field against the solar wind, Mars then lost the light gases such as water vapor, nitrogen, and oxygen over time. Solar energy could not be stored effectively anymore in the atmosphere, and Mars cooled as a result. However, in the first $1-2$ billion years after Mars's formation, conditions may have been quite favorable for the evolution of Life, so that

Fig. 9.36:   **Top:** Image of the planet Mars created from 100 individual images taken by the Viking-1 orbiter. It shows the Grand Valley, which is about 3000 km long, up to 600 km wide, and 8 km deep. **Bottom:** Martian landscape with the approx. 30 − 40 meter high Twin Peaks in the background. The rocks in the foreground have sizes of 5 − 10 cm. The marked rock formation is about 70 cm wide and 50 cm high and is about 6.40 m away. Image taken by the Mars Pathfinder July 1997. [credit: NASA]

early organisms even had the chance to reach higher evolutionary stages.

Exploration of Mars and his surface through space missions began in 1960. Since then, there have been a total of 49 attempts to send research satellites to Mars. The success rate in the years from 1960 to 2000 was memorably low — 32 attempts but only 10 successes or partial successes. The first successful fly-by was accomplished by NASA in 1965 with the Mariner-4 spacecraft. In 1971, Soviet scientists accomplished the first landing of a module with Mars-3, but contact broke off after less than 2 minutes, without it having sent any pictures. With Viking-1 and Viking-2 NASA succeeded in 1976 in making a soft landing. The modules sent data and analyzed soil samples for almost 5 years.

Between 2000–2020, there were 17 Mars missions, 14 of which were successful. Mars mobiles (called rovers) were used for the first time. Water was found in the Martian soil as well as evidence of larger water deposits. Even the early history of Mars could roughly be traced from the rock samples.

During the writing of this book, in February 2021, the mobile module "Perseverance" was successfully launched to Mars. With it on board was the helicopter "Ingenuity". A wide and varied field of tasks will have to be worked through in the coming years with the goal of illuminating the planet's past, finding traces of past Life, and perhaps even encountering primitive life forms, i.e., those that have been able to adapt to the increasingly hostile conditions on the planet Mars.

Another celestial body that is also becoming the focus of current research is Saturn's moon Enceladus. The Cassini spacecraft, launched in October 1997 by NASA in collaboration with the European Space Agency ESA and the Agenzia Spaziale Italiana ASI reached Saturn after a nearly 7-year journey. The first highlight of the mission was the soft landing of its Huygens module in January 2005 on Saturn's largest moon Titan, which is covered with a dense atmosphere. This landing is the only one accomplished so far on a celestial object in the outer solar system (i.e., beyond Jupiter). An important fact one may want to keep in mind is that a radio signal will travel more than 1 hour per pass. A quick intervention to avoid a dangerous situation is therefore impossible. The Huygens module was able to transmit data about Titan for about 3 hours with its 10 watt transmitter.

Cassini remained in Saturn's orbit until 2009, exploring its moons, and — with the discovery of another 20 moons — it brought the total count to 82. In 2008 the probe finally passed the moon Enceladus at a distance of only 25 kilometers and uncovered some astounding properties of this object.

Enceladus is an ice moon, which is about 504 km in size and thus sixth in the present size-ranking. It has an internal, unknown heat source that leads to eruptions on its surface, which are similar to geysers on Earth in areas of increased volcanic activity (see Fig. 9.37). On Earth, very distinct biotopes develop in such hydrothermal vents, as they have been discovered e.g., on the ocean floors. Of these, the best known hydrothermal source is the volcanically active mid-Atlantic ridge, where highly unusual organisms have been found. They obtain their energy by oxidation of the escaping hydrogen sulfide and splitting off molecular hydrogen. It is therefore conceivable that similar biotopes exist on Enceladus close to these eruptions. This argument is supported by the fact that the gases expelled by such organisms, such as molecular hydrogen and methane, have been detected in the eruptive gas fumes of Enceladus.

The origin of the heat generation in Enceladus is, however, unclear. Tidal effects could well be responsible. They are caused by the relative proximity to its parent planet (semi-axis: $\approx 238,000$ km, radius Saturn: $\approx 120,000$ km) and its high orbital velocity (≈ 12.6 km/s or $\approx 45,360$ km/h) as well as to the gravitational and partly resonant interaction with its neighbors. These create stresses in the interior of the planet, which, in case of sudden discharge, lead to extreme heat generation.

Two close-up images of the surface of Enceladus are shown in Figs. 9.38 and 9.39.

Fig. 9.37: **Top:** The surface of Saturn's moon Enceladus. Enceladus has a diameter of about 504 km and is mainly composed of water ice. **Bottom:** Gas eruptions discovered by the Cassini spacecraft, which are produced by hydrothermal vents in the interior of this celestial body, could prove to be biotopes for living organisms. [Credit: NASA]

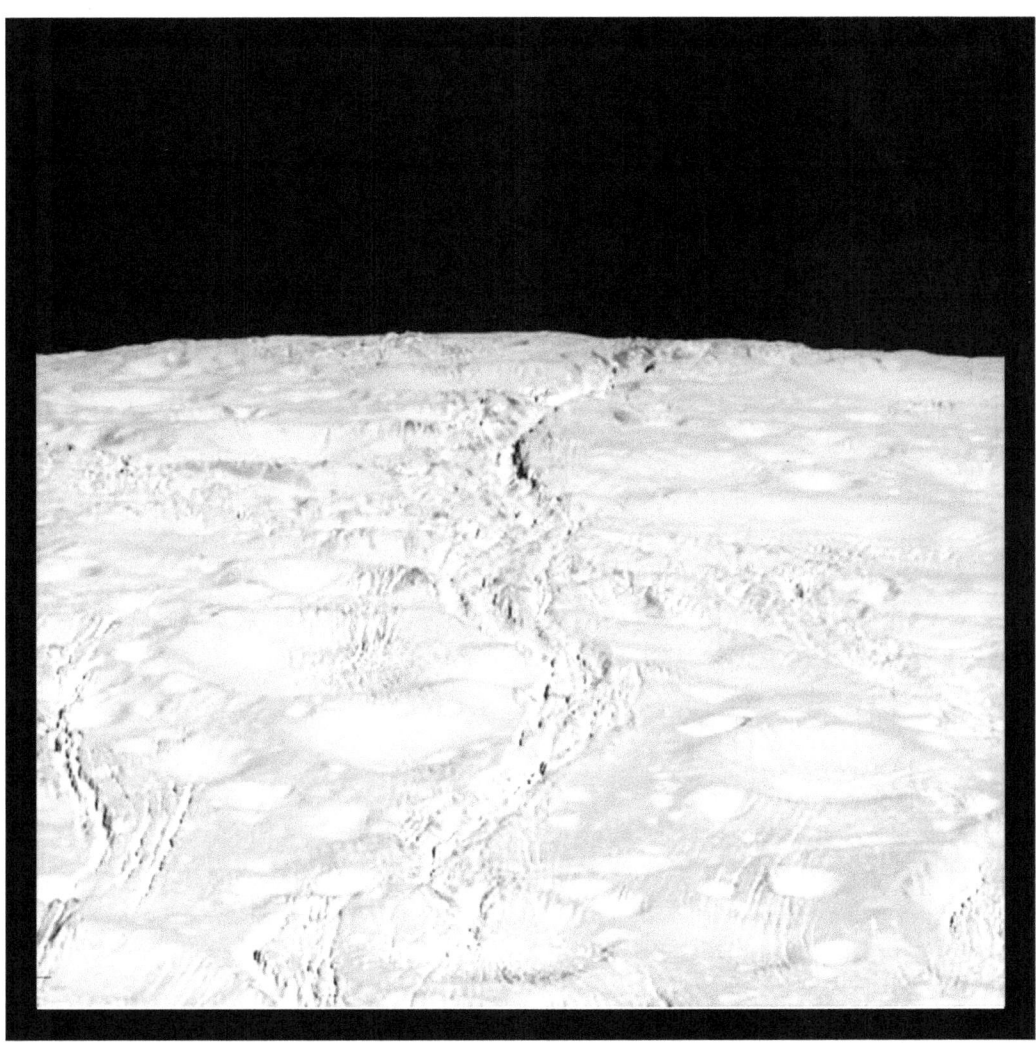

Fig. 9.38:   Photograph of the surface of Saturn's moon Enceladus, here the north pole region. The image was taken during one of the Cassini orbiter fly-bys. [credit : NASA]

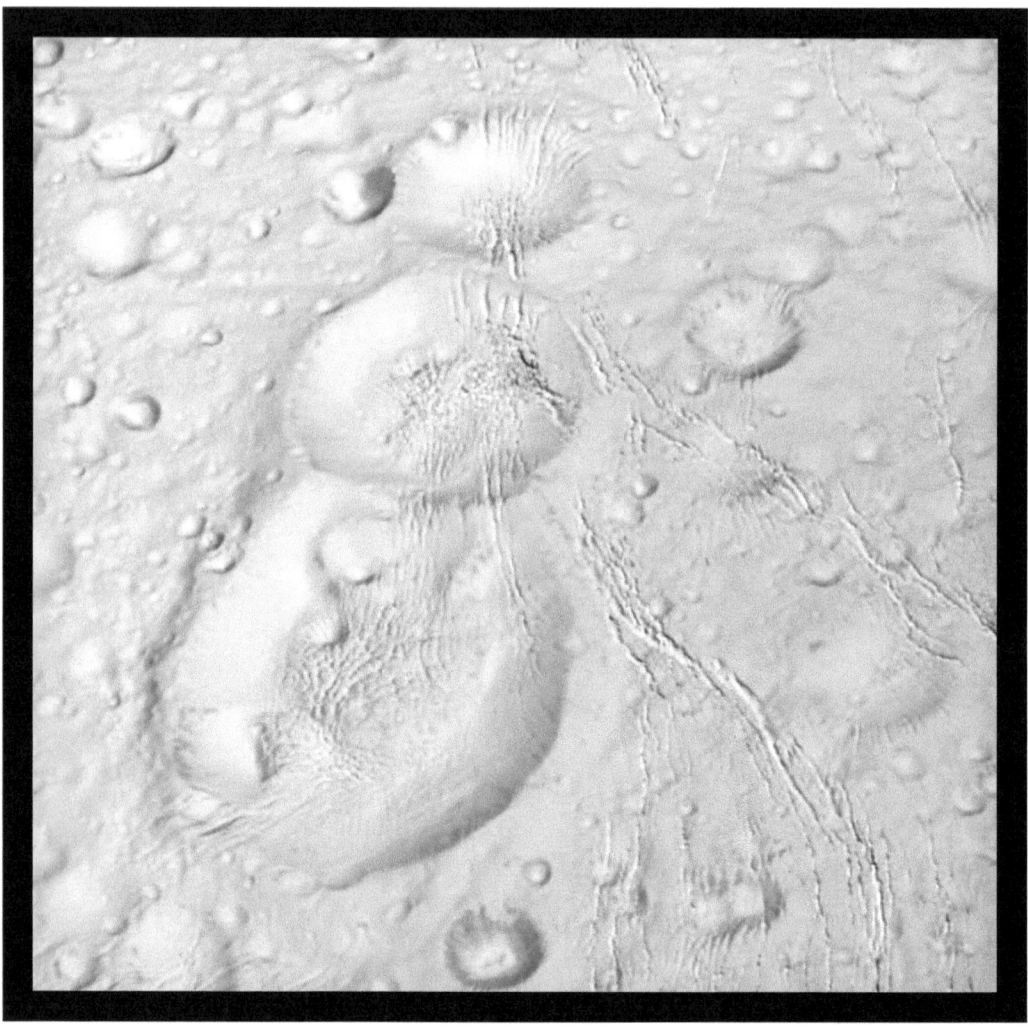

Fig. 9.39: Photograph of the surface of Saturn's moon Enceladus at close distance showing a snowman structure in the image. The image was taken during the Cassini orbiter fly-by at an altitude of about 25 km. [credit: NASA]

Chapter 10

Relativity

The theory of the relativity of time is in many respects not intuitive at all. It seems totally at odds with our daily experiences. Indeed, it is not immediately obvious

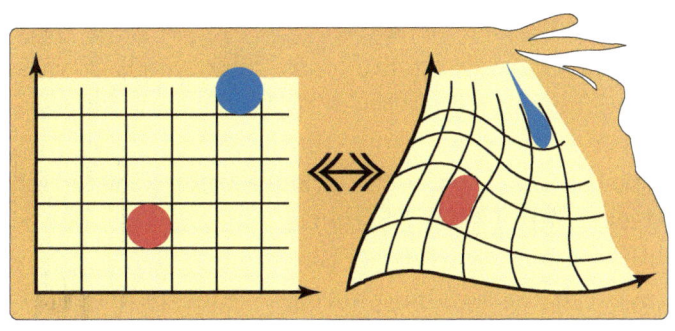

why velocities do not simply add up; why one ages slower when moving (but only when "someone else" compares the elapsed times); why gravitational forces and thus also inertial forces (e.g., acceleration and centrifugal forces) change time; why massless light is subject to gravity; why the head ages faster than the feet; and how, in fact, the formula $E = mc^2$ comes about. Yet, the theory of relativity has long since found its way into the normality of daily life, namely whenever time sequences at different positions have to be adjusted with high accuracy. This applies, for example, to the Global Positioning System (GPS) in navigation or to digital radio traffic (cell phones), which requires a reliable time standard. Here, differences in the mass distributions of the Earth and the resulting differences in gravity, which are far below human perception, produce already unacceptable time differences. If these effects were not taken into account, airplanes would not find their destination, the navigation system in a car would be useless and cell-phone conversations would be lost in a clutter of voices.

All these effects follow solely from the constancy of the speed of light "c", and over time, this constancy has been determined experimentally with an ever increasing accuracy. Interestingly, this peculiar feature already followed from Maxwell's equations of electrodynamics and was known even at the

D. Frekers, P. Biermann, *Universe, Neutrinos, Stars and Life*, https://doi.org/10.1007/978-3-662-70729-6_10

time before Einstein, however, without the consequences being recognized. Today, the more pertinent questions are, what mechanism is responsible for this constancy and at what stage of the Early Universe did this come into effect. On the agenda of the questions to be solved in the future, this one is placed right at the top.

The theory of relativity has two basic pillars, which are named theory of "Special Relativity" and theory "General Relativity". The theory of Special Relativity is mathematically simple, but only applicable in a meaningful way for very "special" cases. Its range of validity is limited to the case of vanishing masses, vanishing forces and Euclidean geometry. Yet, already the basic notions of "time dilation" and "length contraction" appear, and also the formula $E = mc^2$ can be developed with some lines of easy mathematics. General Relativity is mathematically demanding, but capable of describing all known relativistic phenomena with impressive accuracy, taking into account masses and forces. This is true up to extreme masses as they occur in neutron stars or Black Holes.

10.1   Theory of Special Relativity and highlights

Special Relativity only knows one variable, which is the velocity v or its value $\beta = v/c$ as the fraction of the speed of light c. It is the parameter, which appears in virtually all formulas and most notably in the form derived already in 1892 by H. Lorentz [1] (also called Lorentz factor):

$$\gamma = \frac{1}{\sqrt{1-\beta^2}}$$

Relevant values given in the Table 10.1 may help estimate certain relativistic effects, where, for simplicity, $c = 300,000\,\text{km/s}$ has been used for the speed of light.

---

[1] Hendrik Antoon Lorentz (1853 – 1928), Theoretical physicist and professor at Leiden University (NL), Nobel Prize in Physics 1902.

Tab. 10.1: Some key relativistic numbers as a function of velocity. The factor γ determines length contraction, time dilation and inertial mass increase of a uniformly moving object.

velocity (km/s)	$\beta = \mathbf{v}/\mathbf{c}$	γ
100	$3.33 \times 10^{-4} = 0.033\ \%$	1.000 000 06
300	$1 \times 10^{-3} = 0.1\ \%$	1.000 000 50
1000	$3.33 \times 10^{-3} = 0.333\ \%$	1.000 005 56
30,000	$0.1 = 10\ \%$	1.005
100,000	$0.33 = 33\ \%$	1.061
270,000	$0.9 = 90\ \%$	2.294
297,000	$0.99 = 99\ \%$	7.089
299,970	$0.999\ 9 = 99.99\ \%$	70.71
299,999	$0.999\ 996\ 66 = 99.999\ 67\ \%$	387.3
299,999.999	$0.999\ 999\ 997 = 99.999\ 999\ 7\ \%$	12,247.4
300,000	1	∞

There is a variety of thought experiments dealing with travel at relativistic velocities to show how to bridge large distances. Such thought experiments are useful at best to approach the field of the theory of relativity and to deal with physical features, but they are in most cases unrealistic. In order to bring a hypothetical spaceship to 90% of the speed of light and thus shorten the time span in this spaceship, which is still very modest at $\gamma = 2.29$ (see Tab. 10.1), one quickly arrives at required energies that are completely beyond reality (i.e., by including masses, supply, fuel, efficiency factors and of course also by considering tolerable acceleration forces during the outward and return journey). Such energies easily exceed the present annual energy consumption on Earth by a factor of 1,000 to 100,000, whereby the question of which fuel to use and how to produce it, still remains unanswered.

The movie and TV series "Star Trek" used antimatter as a solution. However, this cannot be easily purchased in large quantities (kilo-tons are needed!!).

Note: *To get an idea of the energies, it may be useful to turn back to the subchapter 3.2. There it was described how destructive meteorites are if they hit Earth with only about 30 – 50 km/s. Indeed, the masses involved there are much larger than those of conceivable spaceships, but note that the energy scales with v^2 (i.e., $E \propto v^2$) and thus a factor $10^7 - 10^8$ comes into play with spaceships!*

Muons from cosmic rays

In cosmic rays there is a component that impressively confirms the theory of Special Relativity. The interaction of high-energy cosmic rays with the atmosphere produces a large number of short-lived particles, including muons. Muons are the heavy partners of electrons. They have about 205 times the mass of electrons and suffer little deceleration as they pass through the atmosphere. However, their lifetime is only about $2.2\,\mu s$. They are generated primarily at about $30-50\,km$ altitude. To reach the Earth's surface from this height, light needs already $100-170\,\mu s$, i.e., about $50-75$ times the lifetime of a muon. Nevertheless, high-energy, and thus relativistic muons reach the Earth's surface in large numbers. In accordance with Special Relativity, the time-dilation factor γ occurs here, extending their lifetime for an outside observer by at least this factor, $50-75$. From the Table 10.1, one sees that the muons thus had a velocity of at least 99.99% the speed of light, or an energy greater than $5-7\,GeV$. And indeed one finds that the spectrum of muons from these cosmic-ray interactions goes even far beyond these energies.

But how does this journey look from the muon's point of view? The muon doesn't know anything about the time lengthening and doesn't see itself in motion either. It has its own time, called "proper time", and this says that it must decay after about $2.2\,\mu s$. Therefore it sees the Earth with the same time dilation factor γ coming towards itself. But since there should be only one observer, namely the one who makes the measurement on the muon, he/she will conclude that the muon in its own muon system sees the distance to the Earth shortened by the same length contraction factor γ (because this is the only way to overcome the distance to the Earth's surface in $2.2\,\mu s$). Thus, moving systems possess intrinsic times, which are not congruent with each other. Time intervals in two different systems are relative and depend on how fast the two systems move with respect to each other.

In the limiting case of the speed of light, the proper time is "zero", i.e., time "stands still", and a decay (if it had been possible) does not take place for our observer anymore. In the same way the length contraction leads to a zero length, i.e., finite distances do not exist anymore either. These are exactly the properties of light: it has a proper time "zero", and distances "A-to-B" in the light system disappear.

Neutrinos from supernova SN1987 A

On February 23, 1987, at about 7:35 h (Universal Time), neutrino detectors located at various locations on Earth registered the neutrino signal from supernova SN1987 A, as it collapsed into a neutron star. The explosion occurred in the Large Magellanic Cloud, and neutrinos sent from there to Earth had thus traveled 163,000 light-years.

The registered energies of the neutrinos were between 10 and 40 MeV. Assuming for the neutrino mass the present upper limit of $m_\nu c^2 \approx 0.1$ eV determined by the Planck experiment, and an energy of 40 MeV (thus $\gamma = 4 \cdot 10^8$), then it took a neutrino in its proper time — following the theory of relativity — just $3\frac{1}{2}$ hours for this journey. But unfortunately, there were no clocks on board, which could have been read out for verification.

Nonetheless, this directly observed supernova event is worth mentioning for a few other reasons.

The number of neutrino events registered in all detectors together was quite modest at 21, and interestingly, from this number the total number of emitted neutrinos could be determined to be about 10^{57}, in agreement with theoretical predictions of supernova explosions.

However, there was a rather unfortunate experimental shortcoming at that time. The various detectors were not properly synchronized with each other — in 1987 there was no GPS yet — so that an exact determination of the time sequences of the neutrino events was not possible. Only the Brookhaven neutrino detector was able to provide relative time stamps for the 8 neutrino events it registered. It was found that the highest-energy and supposedly fastest neutrino of 38 MeV reached the detector 5 seconds earlier than the lowest-energy one of 22 MeV. Assuming that both neutrinos had been created at the same time, the 5 seconds flight time difference was used to infer a neutrino mass of 37 eV. Of course, this result sounds a bit like playing with numbers, but on the other hand, the value is not completely absurd.

From this result one may also conclude that supernovae are not the best choices for determining the neutrino mass. In the example quoted above, the flight time difference of the two neutrinos with mass of $m_\nu c^2 \leq 0.1$ eV (i.e., the present Planck limit) would be less than $36\,\mu s$.

A fact to note: The neutrinos from supernova SN1987 A reached the Earth about the same time as the light from the explosion. This implies that although the neutrino energy is about $6-7$ orders of magnitude higher than the energy of, say visible light, and despite their tiny neutrino mass, the relativistic velocity limit "c" also applied to them — a far from trivial observation. In fact, without such a limit the neutrinos would have been about 28,000 times faster than light and would have needed just $5\frac{3}{4}$ years for their travel to Earth. They would have reached Earth already 163,000 years ago.

A legitimate and unanswered question therefore remains:

> **Who tells the neutrino not to be faster than light, and anyway, what does the neutrino have to do with light in the first place ?**

10.2   Theory of General Relativity and highlights

Travelling and ageing

The twin paradox is certainly the most frequently cited phenomenon of the theory of relativity: Twin B leaves twin A and goes on a relativistic journey. Upon his return he finds twin A as an old man, while he has remained in youthful age. — In 1971 Joseph Hafele and Richard Keating got to the bottom of this phenomenon by an experiment [2]. Four cesium atomic clocks were brought aboard a commercial airliner. They were then flown once eastward (total time: 65.4 hours, flight time: 41.2 hours) and once westward (total time: 80.43 hours, flight time: 48.6 hours) around the Earth via several stations (see A and B in Fig. 10.1). Finally, back at the

---

[2] J. Hafele und R. Keating, *Around the world atomic clocks: predicted relativistic time gains*, Science 177, 166 (1972).

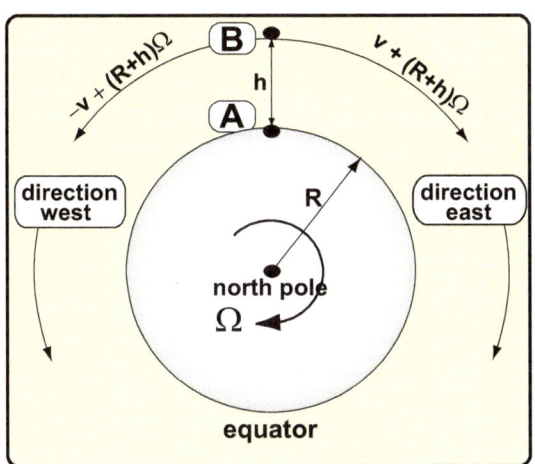

Fig. 10.1: When traveling by air eastward at an altitude h, one has to add to the cruising velocity v of traveller B the velocity of the Earth-bound observer A caused by the Earth's rotation, which is $(R + h) \cdot \Omega$. When traveling westward, the travel velocity is opposite to the Earth's rotation and therefore lower with respect to point A. Thus, due to the theory of Special Relativity, "time clicks" are reduced for eastward travel, i.e., ageing is slower eastward. The effect has been measured for the first time in 1971 using several high-precision atomic clocks and comparing the time of the Earth-bound observer A with the one of the traveler B after a trip around the Earth, once eastward and once westward. The flight times took about $40 - 50$ hours each.

starting point A, these clocks were compared with the atomic clock of the United States Naval Observatory. During the flights, all flight data, such as speed, altitude and position were continuously recorded, including the inaccuracies, which were still common at that time, and which in the end were reflected in the total error of the values calculated by the theory of relativity. The comparison of the clocks with the clock at the starting point A showed that clock B had lost (-59 ± 10)ns on the journey to the east (i.e., had remained 59 ns younger), while it had gained (273 ± 7)ns on the journey to the west (i.e., had become 273 ns older). According to the theory of relativity, this was to be compared with (-40 ± 23)ns and (275 ± 21)ns !!

The situation is explained in Fig. 10.1, where one may recall that the speed of the Earth's rotation was always larger than the speed of airplanes during the flights.

Meanwhile, there are a number of new editions of this experiment, also under controlled laboratory conditions. By now it has even been possible to determine relativistic time dilation at velocities below $10\,\text{m/s}$ and thus in the order of 10^{-16}, in an impressive agreement with Relativity.

It has certainly not escaped the attention of the attentive reader that the experiment described by Hafele and Keating does not adhere to the basic idea of Special Relativity. In fact, in this experiment there is no inertial system, i.e., a system free from external forces. The clocks in that experiment are in different gravitational potentials, they go through different acceleration phases and experience centrifugal forces due to the rotation of the Earth. All of these things require a much more elaborate set of calculations within the framework of the theory of General Relativity, which with increasing experimental accuracies is, of course, mandatory.

According to General Relativity, clocks in the higher gravitational potential, i.e., at higher altitudes, go faster compared to those in a lower gravitational potential near the Earth's surface. In the Hafele-Keating experiment, this effect was the only one that conformed to General Relativity when calculating the time differences. It amounted to $(+144 \pm 14)$ns, for the eastward flight. If it had not been taken into account, the measured time gain on this flight instead of (-59 ± 10)ns would have been (-243 ± 18)ns, and thus been significantly larger. All other correction effects were in the range of the error margin, as later calculations revealed.

Head and foot, mountain and valley

The fact that clocks run faster in the higher gravitational potential holds a number of amusing effects, but one should not get carried away, as these have at the same time important technical consequences.

In the gravitational field near the Earth's surface, the relative rate difference of two clocks is given by

$$\frac{\Delta t}{t} \approx \frac{g_{\text{Earth}} \cdot \Delta h}{c^2}$$

where g_{Earth} is the acceleration due to gravity $\approx 9.81\,\text{m/s}^2$ and Δh is the height difference of the two clocks.

The measured gravitational time change effect conforms with Relativity [3] even when measured at 30 cm height difference, where it amounts to 3×10^{-17} according to the above formula. If one extrapolates this value to the age of the Earth of approximately 4 billion years, then just about 4 seconds would have accumulated.

A human who lives in the mountains 1000 m above sea level is, when assuming a life span of 70 years, around 0.24 ms "more aged" than his twin sibling, who lives at the coast. However, both have to deal with the problem that their heads are approx. $0.2\,\mu$s older than their feet (body height about 175 cm, times of sleeping when lying flat deducted).

Conversely, these examples say that time is different at every point in the Universe, i.e., it is relative!

This is an extremely important realization and finds application in the technical area with the conception of the Global-Positioning-System (GPS). In order to make sure that the clocks in the satellites that orbit around Earth at an altitude of about 20,000 km do not run out of synchronization with the Earth-bound control clock, the frequencies of the orbit clocks are reduced by exactly this relativistic contribution.

Mercury perihelion movement

As we have seen, a mass changes the time sequences with respect to an observer by its gravity, but not only that, it also changes the metric (or geometry) of space in its immediate vicinity. Astronomers first encountered this phenomenon as early as 1855 in the course of a precise survey of Mercury's highly eccentric orbit around the Sun. The French astronomer and mathematician U. J. J. Le Verrier (known for the discovery of the planet Neptune, whose existence he deduced in 1846 on the basis of orbital perturbations of Uranus and whose exact determination of its position

---

[3] C. W. Chou, et al., *Optical Clocks and Relativity*, Science 329, 1630 (2010).

Perihelion rotation per century			
Planet	total	General Relativity	observed
Mercury	572.4″	42.92″	43.11″±0.45″
Venus	18″	8.6″	8.4″±4.8″
Earth	1163″	3.8″	5.0″±1.2″
Mars	1598″	1.4″	1.5″±0.15″

Tab. 10.2: Perihelion motions of the inner planets of the solar system in units of arcseconds per 100 years. After subtracting the effects of the planets on each other, the remaining part is correctly described by General Relativity.

eventually led to the discovery) was able to determine the perihelion rotation of Mercury's orbit to about 570 arcseconds ($570″\simeq 0.16°$) per century from precise measurements of Mercury's transits in front of the solar disk. — (**Note:** *These Mercury transits occur reasonably regularly at intervals of* $3\frac{1}{2}$, *6, 7, 9 or 11 years, always in May or November.*) After subtracting the contributions caused by all other planets, which are mainly those caused by Venus (due to proximity) and Jupiter (due to mass), a discrepancy to the measured value of about 43″ per century remained. Of course, it goes without saying that the calculations assumed classical Newtonian mechanics. The only plausible explanation at that time was seen in the presence of another inner planet. But this planet was never found, and only 60 years later was General Relativity able to solve this riddle:

The additional perihelion rotation of Mercury of about 43 arcseconds per century is due to the curvature of space generated by the gravitational mass of the Sun.

Figure 10.2 shows Mercury's rosette-like orbit and the resulting perihelion motion in a highly exaggerated form. Figure 10.3 on the left shows a photographic image of Mercury just before it reaches the solar disk, and on the right it shows its positions in front of the solar disk over the course of the next approximately $5\frac{1}{2}$ hours. The event occurred on November 11, 2019, and is one of the very rare events when Mercury's orbit passes close to the center of the solar disk (central transit) (here to within 76″). The next even more central transit will occur on November 12, 2190. This will also be the transit closest to the center of the solar disk in this millennium with a minimum distance to the center of the solar disk of only 9″. The sidereal (i.e., related to a distant fixed point) perihelion rotations of all inner planets have by now been accurately measured, and are summarized in Table 10.2.

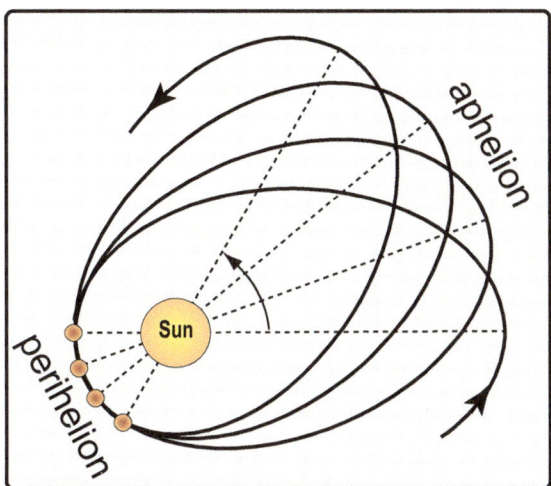

Fig. 10.2: Perihelion motion of a celestial body in case of an eccentric orbit around the Sun. In this representation, the effect is grossly magnified because in the case of Mercury the angle of rotation is only $572''$ ($\simeq 0.16°$) per century. Due to its proximity to the Sun and its particularly pronounced eccentric orbit, this value includes $43''$ ($\simeq 0.012°$) per century as a result of space curvature. Mercury thus shows the largest relativistic effect among the planets. The numerical value results from the theory of General Relativity and eliminates the discrepancy between the measured value and classical Newtonian theory that existed since about 1855.

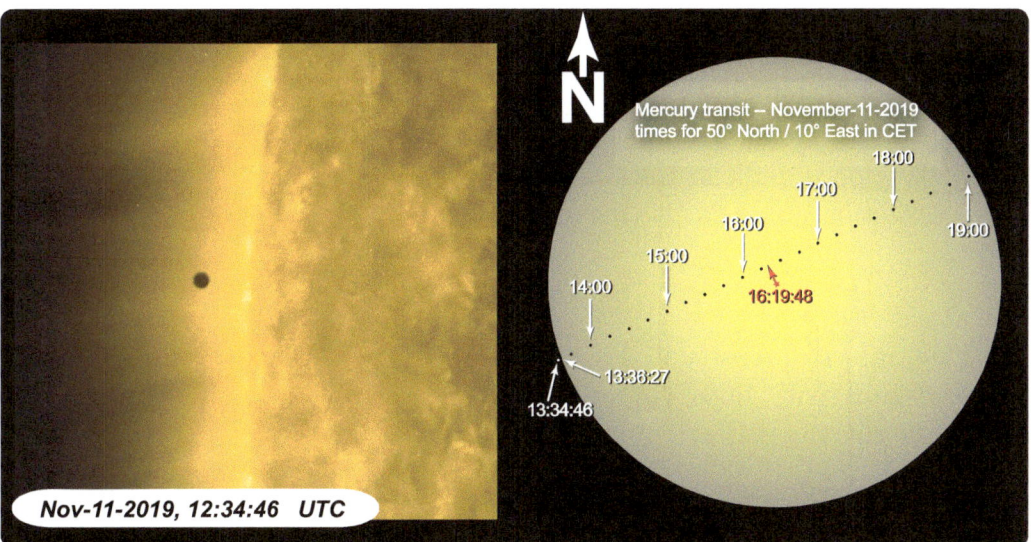

Fig. 10.3: **Left:** The planet Mercury shortly before it reaches the solar disk. The image was taken on Nov-11-2019 at 12:34:36 Universal Time Coordinated (UTC), or 13:34:36 Central European Time (CET). **Right:** The transit of Mercury in front of the solar disk. The individual times are indicated. The smallest distance to the center of the solar disk was reached at 16:19:48. The transit lasted just under $5\frac{1}{2}$ hours. From the transit velocity the perihelion position of Mercury can be determined. [Figure left: NASA/SDO/AIA, figure right: edited after the original and with permission of *Arnold Barmettler, www.CalSky.com*.]

Light deflection in a gravitational field

The fact that light is deflected in the gravitational field of a mass was already mentioned in chapter 6.2 on pages 60ff. Arthur Eddington had measured this phenomenon in the course of a total solar eclipse on May 29, 1919 (see also Fig. 6.5). The experimental proof of such a space curvature understandably caused some excitement, and not only among scientists. After all, General Relativity was by no means an established theory at that time, nor was its understanding intuitive at all.

In retrospect, it is certainly useful to recall the peculiarity of the deflection of light in the gravitational field of the Sun in order to understand the associated amazement at the experimental result. Here are three theoretical models and their statements, whereby the Fig. 10.4 shall serve as explanation: ($G \Rightarrow$ gravitational constant, $M \Rightarrow$ solar mass, $r_{min} \Rightarrow$ smallest distance of the light trajectory to the Sun)

▶ **Classical Newtonian theory:** Assigning a mass to light according to $E = mc^2$, and therefore $m = E/c^2$, one can calculate the deflection of light in the classical sense. The deflection angle $\Delta\phi$ is then given by:

$$\Delta\phi = 2 \cdot \frac{GM}{c^2} \cdot \frac{1}{r_{min}}$$

▶ **Special Relativity:** The theory of Special Relativity does not know masses, consequently the deflection angle is:

$$\Delta\phi = 0$$

▶ **General Relativity:** In General Relativity, the curvature of space is added to the classical deflection angle. The slightly more involved calculation gives:

$$\Delta\phi = 4 \cdot \frac{GM}{c^2} \cdot \frac{1}{r_{min}}$$

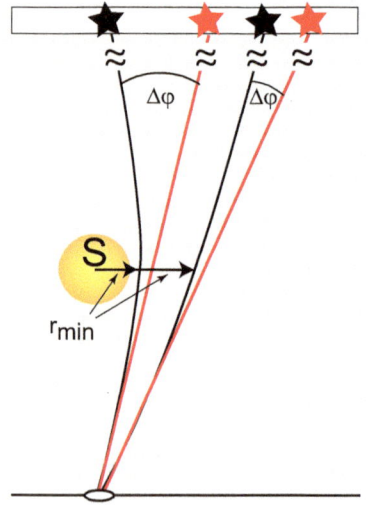

Fig. 10.4: Light deflection in the gravitational field of the Sun according to General Relativity for the example of two stars; black: actual, red: observed positions.

The deflection angle predicted by General Relativity is thus twice as large as the one from classical theory — an effect that cannot be overlooked, even with the tiny absolute magnitude of less than one arcsecond in the case of the Sun. The outcome of the measurements were in convincing agreement with General Relativity, which thus experienced its first major confirmation.

Addendum: On October-8 each year, the quasar "3C279" is eclipsed by the Sun. A measurement of the radio signal deflection by Very Long Baseline Interferometry (VLBI) relative to the nearby quasar "3C273" allows an accuracy of $0.001''$ and serves as a precision test of General Relativity.

The perihelion rotation of Mercury or the deflection of light in the gravitational field of the Sun were tiny effects. Nevertheless, in the following 100 years, they had a decisive influence on a multitude of new technical developments, which meanwhile reach into everyday areas (e.g., mobile phones, GPS, precision time standards)

We have seen how the theory of relativity re-defines the nature of time, how space, which surrounds us as an initially abstract entity, changes and unites with time to form a new quantity, space-time, and what role gravity (or the presence of mass) plays in this. All of this was due to the constancy of the speed of light, which is assumed to be the upper limit for any form of information transfer, be it light, neutrinos or ordinary mass.

In the previous collection of implications resulting from of the theory of relativity, rotating systems have been largely excluded. A rotating system always requires a reference point with respect to which it rotates, but there is no such absolute reference point, which could be identified. This problem had already arisen with Foucault's pendulum $\rightarrow$ the Earth rotates, but with respect to what, and who or what defines the centrifugal forces arising with the rotation? We will see later, in the context of neutron stars and Black Holes, to what extent General Relativity can describe such a situation and what the so-called "co-rotating inertial system" or the "dragging of the inertial frame" is all about.

Images of gravitational lensing

The next pages will primarily be a collection of images (Figs. 10.5 – 10.15) intended to provide some insight into what physics information current and future research can extract from gravitational deflection and gravitational lensing.

Gravitational Lens

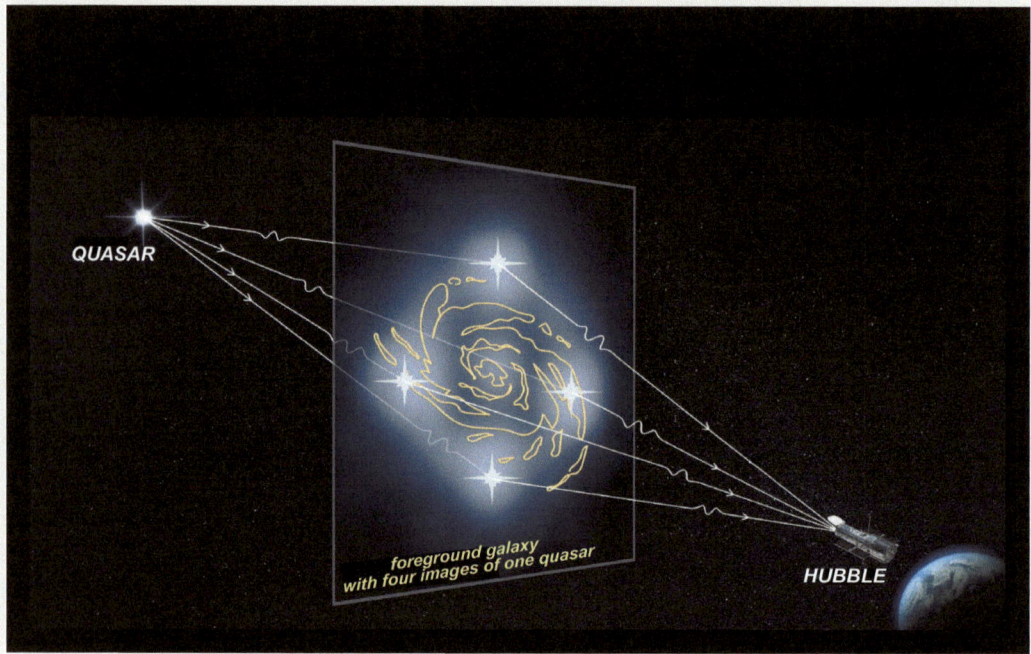

Fig. 10.5: A far distant quasar is imaged onto the Hubble telescope by a foreground galaxy. Due to the focusing, the observed amount of light from the quasar is increased by orders of magnitude, thus allowing a view to extreme distances that would be otherwise only barely possible or not possible at all. Quadruple imaging of a quasar or of a distant galaxy is rare because it requires precise alignment with the line of sight of the imaging galaxy and the observer.

[Public Domain, NASA, ESA and D. Player (STScI) - https://photojournal.jpl.nasa.gov/jpeg/PIA23641.jpg]

Einstein Cross

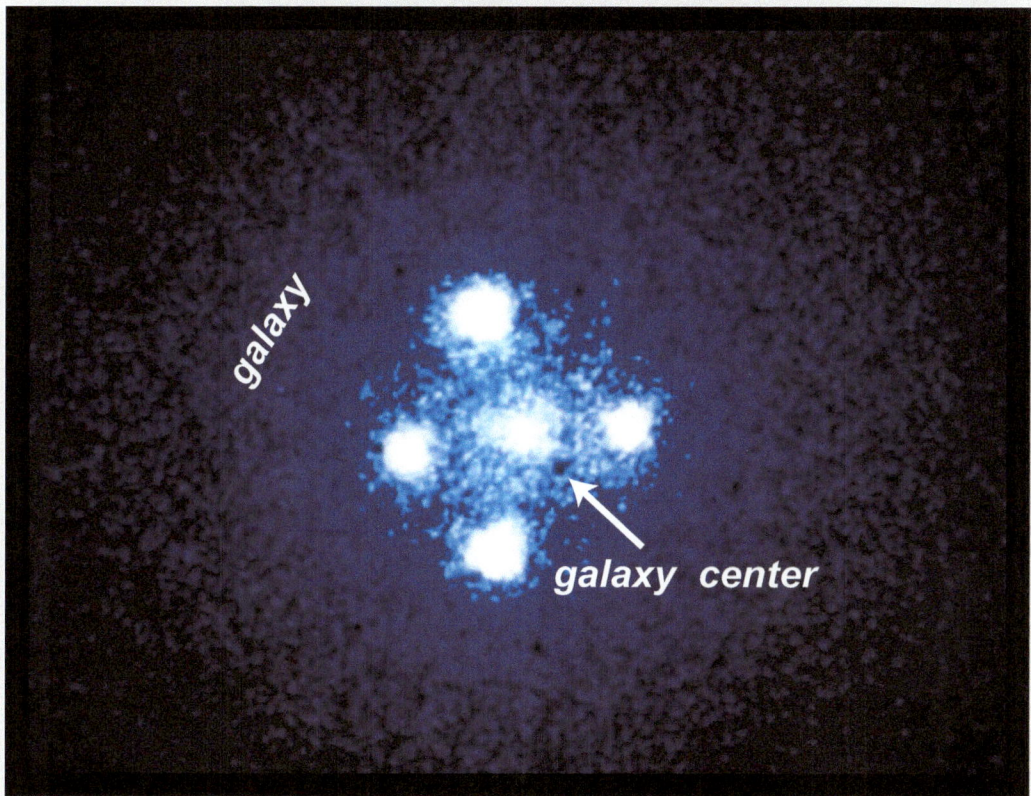

Fig. 10.6: The quasar "QSO 2237+0305" is seen from Earth almost exactly behind the center of a distant galaxy about 400 million light-years away. The quasar is about 8 billion light-years away. The galaxy acts as a gravitational lens, creating four images of the quasar of equal brightness. They appear in the form of a cross (known as the "Einstein cross") with the nucleus of the distant galaxy in the center. This phenomenon was recognized by the fact that the four images have an identical light spectrum. [Credit: NASA, ESA, and STScI]

Einstein Rings

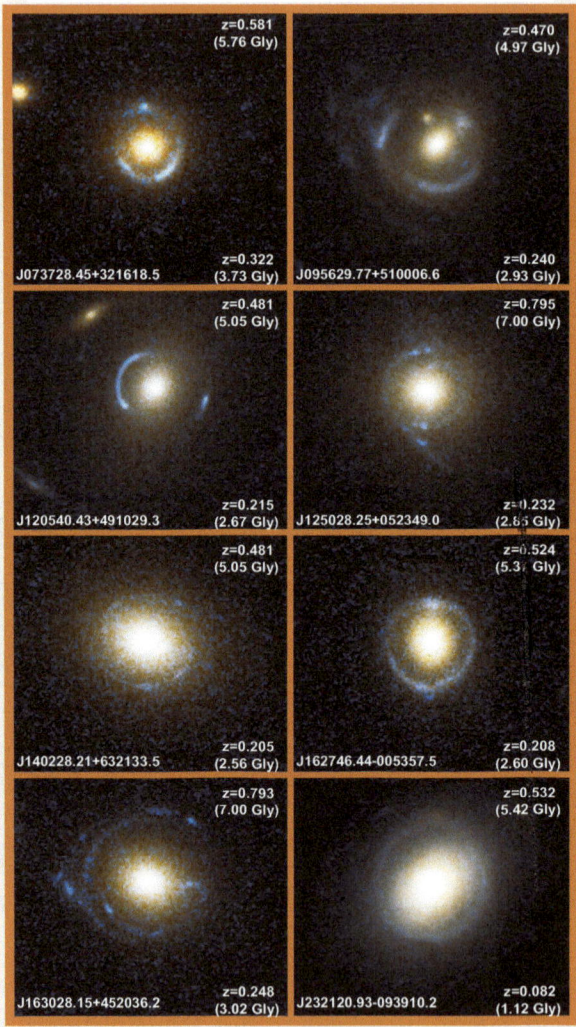

Fig. 10.7: A sequence of so-called Einstein Rings. These are images of background galaxies created by galaxies in the foreground (in the center of each ring), which act as gravitational lenses. The images are from the Sloan Digital Sky Survey (SDSS) project. The astronomical identifiers are given on the lower left, the redshifts z and the distances of the foreground galaxies are given on the lower right. The upper right shows the same parameters of the background galaxies.

[Credit: NASA/ESA, A. Bolton (Havard-Smithonian CfA), SLAC Teams, Photo: STScl-PRC05-32]

Horseshoe Image

Fig. 10.8: A nearly perfect Einstein Ring. A background galaxy acquires the image of a horseshoe. [Credit: NASA/ESA]

"Happy Face"

Fig. 10.9: "Smiling galaxy cluster" or "Happy-Face" cluster. — The two "eyes" are particularly luminous galaxies in the cluster and the rings are images from the distant galaxies behind them created by gravitational lensing. The lensing effect makes it possible to explore epochs of the Early Universe that are hardly accessible by any other means. [Public domain, credit: NASA & ESA Acknowledgement: Judy Schmidt (geckzilla.org)]

Distorted Galaxies

Fig. 10.10: Six images of galaxies taken by the NASA/ESA Hubble Space Telescope. They show oddly shaped galaxies with arcs, streaks, and smeared galaxy structures. These are images of some of the most powerful galaxies radiating in the infrared. They are about 10,000 times more luminous than the Milky Way Galaxy. Due to the gravitational lensing effect, structures in these background galaxies can be resolved from about 100 light-years in size. One can here look back about 8 – 11.5 billion years into the past, i.e., at a time when in these extremely active galaxies developed about 10,000 new stars per year (for comparison, presently our galaxy gives birth to about 2 – 3 stars per year). This enormous star birthrate creates dust and gas clouds surrounding these galaxies, obscuring them in the visible light. They are, however, readily visible in the infrared. The images shown are part of a Hubble project to study distant, ultra-luminous galaxies. Images were taken with the Hubble "Wide-Field" infrared camera. [CC BY 4.0]

Lensed Supernova

Fig. 10.11: The image sequence shows a type Ia supernova explosion imaged by a foreground galaxy due to gravitational lensing. The background image was taken at Mount Palomar Observatory in California, the upper left image is from the Sloan Digital Sky Survey (SDSS) project, and the center image with the lensing galaxy (name: J210415.89-062024.7) is from the Hubble telescope. The upper rightmost image (also from the Hubble telescope) shows the four images of the supernova surrounding the lensing galaxy. The supernova has a redshift of $z=0.409$ (distance 4.4 billion light-years), while the galaxy has a redshift of $z=0.2163$ (distance 2.6 billion light-years).[‡] (Supernova-id: iPTF16geu) [CC BY 4.0]

[‡] E. Mörtsell et al., *Lens modelling of the strongly lensed Type Ia supernova iPTF16geu*, arXiv:1907.06609v2 [astro-ph.CO], (2020)

   A. Goobar et al., *iPTF16geu: A multiply imaged, gravitationally lensed type Ia supernova*, Science 356, 291 (2017)

Pregnant Cosmic Serpent

Fig. 10.12:   A cosmic serpent pregnant with stars. The figure shows how a distant background galaxy with a high star-birth rate is imaged by the gravitational lensing effect of a foreground galaxy. Particularly noteworthy is the high magnification factor, which allows identification of individual newly forming stars in the background galaxy.[‡] [Credit: ESA/Hubble, NASA, Antonio Cava]

[‡] A.Cava et al., *The nature of giant clumps in distant galaxies probed by the anatomy of the cosmic snake*, Nature Astronomy 2, 76 (2018).

   D. Meloy Elmegreen et al., *Resolved galaxies in the Hubble Ultra Deep Field: Star formation in disks at high redshift*, The Astrophysical Journal 658:763 (2007).

Dark Matter at Play?

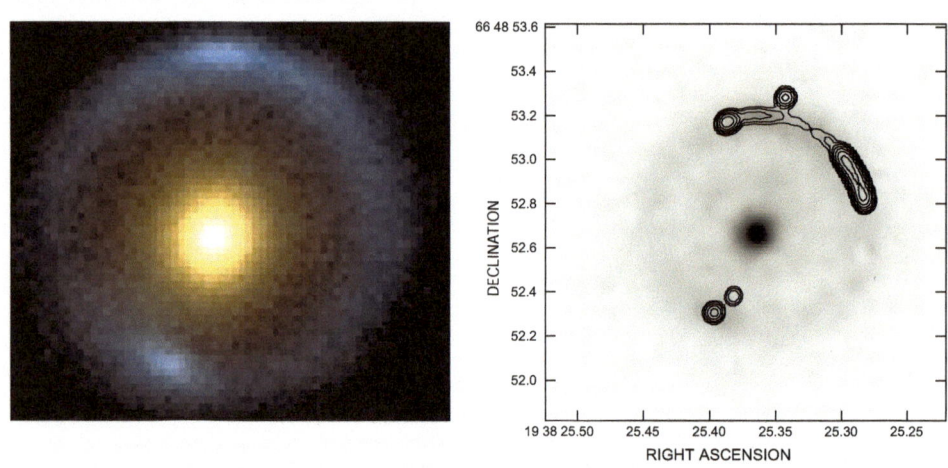

Fig. 10.13: The image on the left shows a background galaxy at a redshift of z=2.059 (distance about 11.05 billion light-years), which is imaged as a nearly perfect Einstein-Ring by a foreground galaxy at z=0.881 (distance about 7.3 billion light-years). The ring shows a weak deformation, possibly caused by a dwarf galaxy in the region of the foreground galaxy. The mass of the dwarf galaxy points to about 190 million suns. However, its luminosity is too small by at least a factor of 3.5 to be consistent with only Sun-like stars. The discrepancy could be a hint that the dwarf galaxy consists predominantly of dark matter, which in turn suggests that dark matter consists of "cold" and "heavy" particles (also referred to as "cold dark matter"). The figure on the right shows an image analysis of the ring.[‡]

[Credit image left: Chris Fassnacht, Univ. of California, Davis; credit image right: Combined HST und MERLIN, L. J. King et al.]

[‡] Vegetti et al. *Gravitational detection of a low-mass dark satellite galaxy at cosmological distance*, Nature 481, 341 (2012). https://doi.org/10.1038/nature10669

Distant Active Galaxies

Fig. 10.14: This sequence of 5 distant galaxies is a composite image using data from the Atacama Large Millimeter/Submillimeter Array (ALMA) and images from the Hubble Telescope. The images created from the ALMA data (in red) show the background galaxies distorted into Einstein-Rings due to the lensing effects caused by the foreground galaxies (in blue). The images in blue were taken with the Hubble telescope.

The ALMA telescope is located on a plateau at an altitude of 5000 meters in the Atacama Desert in Chile. ALMA consists of 66 individual telescopes, which are located in a compound up to 16 kilometers apart. ALMA is currently the largest Earth-based telescope in the millimeter and submillimeter wavelength range. [ALMA (ESO/NRAO/NAOJ), J. Vieira et al.]

Lensed Galaxy

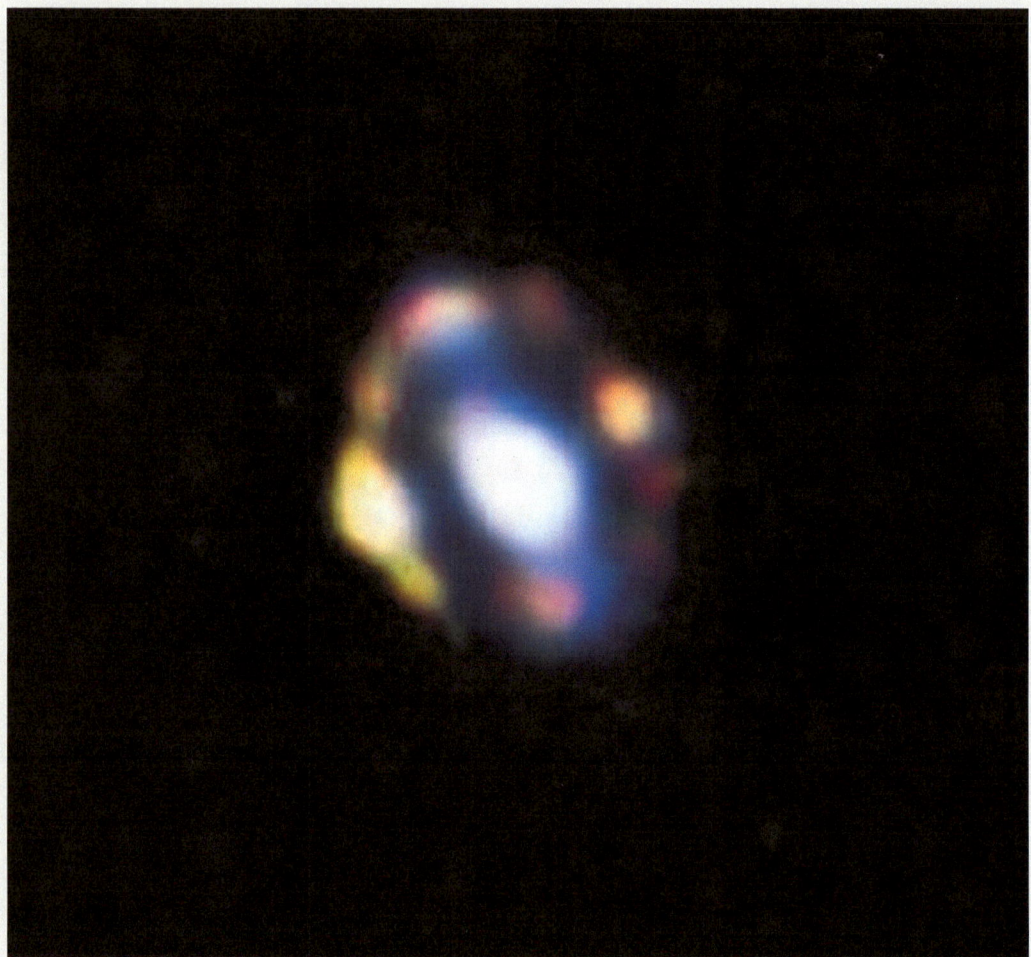

Fig. 10.15: Image of a lensed galaxy taken by the Hubble telescope at a redshift z=1.53 (distance about 9.5 billion light-years). In the center lies the massive core of the lensing foreground galaxy. The light spectrum of the background galaxy provides information about the early and partly violent phase of stellar and galaxy evolution. [CC BY 4.0]

Chapter 11

Neutron Stars, Inertial Systems, Black Holes

Among the known interactions, gravity is the only one that does not seem to resist a transition into an infinity. As a thought experiment, one could compress a mass in such a way that it leads gravity for its part to strive towards infinity as the size of the mass approaches zero. Even General Relativity shows no tendency to stop this unphysical process at a certain limit. General Relativity must therefore be regarded as an incomplete theory, but where and

how does this incompleteness reveal itself? In fact, a genuine singularity (=infinity) of the theory occurs even in a much more realistic scenario, namely at the boundary to a Black Hole, and this happens already at mass densities, which are by no means unusual or exotic. Whereas all other interactions seamlessly merge in the end with quantum mechanics, which prevents such a singular behavior, a protective transition of a similar kind is not in sight for the theory of relativity, at least not for now.

In the following sections, some of the oddities resulting from the theory of relativity shall therefore be touched upon. Astronomical observations and/or particularly precise Earth-based experiments aim to test some of these predictions, also with the high hope of uncovering possible limits of the theory.

11.1   Physics of neutron stars

Some aspects of neutron stars were already pointed out earlier. A neutron star is the leftover core of a supernova explosion and initially consists mainly of iron/nickel elements. The enormous and still increasing gravitational pressure finally presses the electrons into the atomic nuclei and initiates reactions of the form $e^- + p \longrightarrow n + \nu$. The star starts to be "neutronized" and eventually becomes an object consisting almost entirely of neutrons. Gravity is relentless and compresses this structure until about nuclear density (about $10^{14} - 10^{15} \, \text{g/cm}^3$) is reached. At this point, the largely incompressible nuclear matter resists further compression. An equilibrium is reached and a stable neutron star is formed.

Depending on the mass of the original star (about $10 - 20$ solar masses) the neutron star holds between 1.4 and 3.2 solar masses at a radius of only about $10 - 15 \, \text{km}$. The rest has expanded into interstellar space as a result of the explosion. If the original stellar mass was larger than about $20 - 25$ solar masses and if the mass of the neutron star is thus larger than about 3.2 solar masses, a Black Hole is formed. Further compression is not needed for this.

The values given here are approximate, since much depends on how far nuclear matter can be compressed. Not much is known about this nuclear property, which also goes under the name "nuclear compressibility". It constitutes an important research topic in cosmology and astrophysics but no less an important topic in elementary particle physics (e.g., at the CERN research facility near Geneva). The compressibility ultimately decides at which mass-radius ratio the transition of a neutron star to a Black Hole takes place but also what the maximum (or cutoff) frequency is for a rotating neutron star.

In Table 11.1 some characteristic parameters of typical neutron stars are given. One may note that neutron stars still contain about 10% protons, which is a consequence of protons being lighter than neutrons.

Tab. 11.1: Parameters of neutron stars.

Properties of neutron stars	
mass	$1.4 - 3.2$ solar masses
radius	$\approx 10 - 15\,\text{km}$
density (in interior)	$\approx 10^{14} - 10^{15}\,\text{g/cm}^3$
acceleration due to gravity	$\approx (0.2 - 2) \times 10^{11} \cdot g_{\text{Earth}}$
weight of a mosquito	$\approx 50 - 500$ tons
rotational frequency	milliseconds – seconds
record holder for frequency	"PSR J1748-2446ad": $1.4\,\text{ms}$ or $716\,\text{Hz}$
magnetic field	$\approx 5 - 500$ Mega-Tesla
surface temperature	$\approx 10 - 500$ Mega-K
interior temperature	up to $10^{11}\,\text{K}$
composition (surface)	heavy, neutron-rich nuclei
composition (interiors)	neutrons, $\approx 10\%$ protons, free quarks

Some of the parameters listed in the table are particularly worth mentioning, partly because of their sheer size, but also to see what the theory of relativity will have to say.

How many are there?

Since neutron stars arise from supernova explosions, and these are quite frequent events on cosmic time scales, it is estimated that there are about $10^8 - 10^9$ neutron stars in the Milky Way alone. This implies that neutron stars may account for about $1 - 2\%$ of the total stellar mass of a galaxy. However, most neutron stars are inconspicuous and not visible even with the best telescopes. Also, many neutron stars are believed to have ended up as "duds" outside the galaxy disk.

One mosquito to 500 elephants

The enormous acceleration due to gravity, as expressed by the g-factor, which can reach more than 10^{11} times the value on Earth, gives a terrestrial mosquito a weight of nearly 500 tons. A free fall from a height of 5 m results in a speed of 3.1 million m/s (or about 11 million km/h). If this "mosquito" hits the Earth's surface with such a speed, an equivalent of about 18% of the annual energy consumption of a country like Germany would be released in the form of destructive energy. But be careful with Relativity in such experiments: the 5 m distance from the neutron star's perspective is not the same 5 m distance from the perspective of an observer on Earth.

Rotating neutron stars

All known neutron stars rotate because they inherited angular momentum from their parent star. In the course of the contraction, the angular momentum had changed only little, which in the end could lead to rotational frequencies of up to 1000 Hz (pirouette effect). The fastest pulsar "PSR J1748-2446ad" has a rotation period of 1.4 ms (716 Hz); the slowest "PSR J1841-0456" needs just under 11.8 seconds for one revolution.

For comparison: The Sun with a radius of about 696,000 km rotates once in about 26 days. If compressed to an object of about 20 km in radius, the rotation period would go down to about 2 ms, or the rotational frequency go up to 500 Hz (the frequencies scale with the squared radii). However, the Sun will not experience this fate.

Pulsars

Rotating neutron stars generally have extreme magnetic fields of up to 500 Mega-Tesla. This causes intense electromagnetic radiation to be emitted, primarily in the radio-wave range and along the poles. If the magnetic field axis is inclined against the rotation axis as shown in Fig. 11.1, a "cosmic beacon" is created, which flashes at regular time intervals, i.e., each time the line of sight is within the emission cone. If one succeeds in resolving the time sequences with an appropriate detector, these electromagnetic pulses then mirror the self-rotation of the neutron star. Neutron stars with

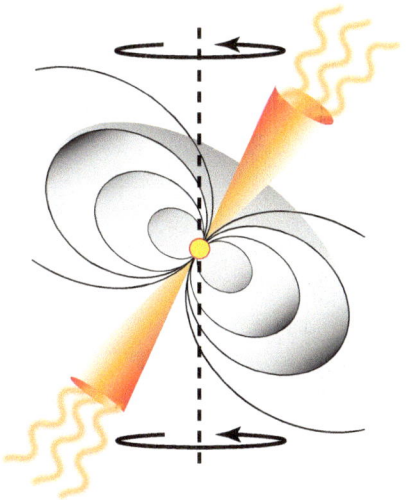

Fig. 11.1: Illustrated is a neutron star (in the center) with its magnetic field, whose axis is inclined with respect to the rotation axis. Along the magnetic field axis, the star emits intense radio radiation that reduces the rotational frequency of the star. The emission cone typically has an angular aperture of about $20°$. Note: In the case of a Black Hole, the magnetic field axis is always aligned with the rotation axis.

these special properties are called pulsars, a word coined as a result of the discovery of the first pulsar in 1967 by Jocelyn Bell and Antony Hewish[1]. By now it is even known that these light pulses have substructures, which indicate structures on the neutron star surface that are smaller than $200\,\mathrm{m}$ in areal size. Radial warps are even smaller than $2\,\mathrm{cm}$. It is therefore not surprising that pulsars are among the most interesting objects of research in astrophysics and cosmology, with many pertinent questions reaching deep into elementary particle physics and the theory of relativity.

"Slow-down" of pulsars

A pulsar gets the energy for the emission of radiation mainly from its own rotation. Due to the co-rotating magnetic field, the charge carriers (electrons) contained in the neutron star's atmosphere are also co-rotating and thereby spiralling along the magnetic field lines. They experience a permanent acceleration, which causes emission of polarized synchrotron

---

[1] Jocelyn Bell was in 1967 at the age of 23 a PhD student at Winston Churchill College, Cambridge University, UK, and Antony Hewish was her PhD supervisor. However, only Antony Hewish (*1924) received the Nobel Prize in 1974 for the discovery of the first pulsar and its correct classification as a neutron star, which he shared with Martin Ryle (1918–1984) for the development of a novel radio telescope system.

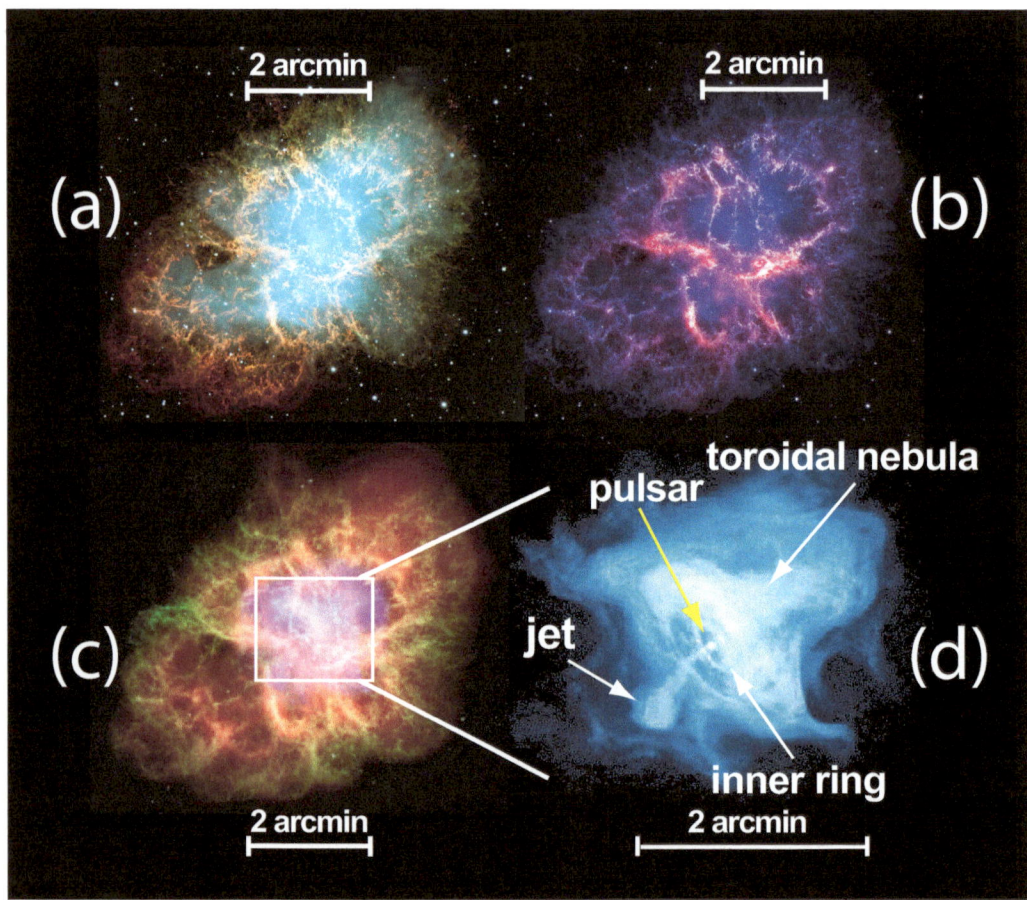

Fig. 11.2: Images of the Crab Nebula (distance about 6200 light-years) in four different wavelength bands:

(a) Superposition of different wavelength ranges; synchrotron radiation (blue-green) surrounded by filamentary structures caused by line emission of atoms (red), here mainly oxygen and silicon.

(b) Radiation in the far infrared.

(c) Composite optical range (red) and X-ray range (blue).

(d) Radiation in the X-ray range created in jets and by shock fronts in the inner ring.

All figures are aligned with each other in the same way. From the large number of spectral investigations carried out over time, it has also been possible to determine the total mass of the original star to be about 10 solar masses.

[Credit: (a) Hubble Telescope NASA/ESA, J. Hester, A. Loll (ASU);   (b) Herschel/Hubble Telescopes, ESA/Herschel/ PACS/MESSTeam, NASA/ESA, A. Loll, J. Hester (ASU); (c) Hubble /Chandra/Spitzer NASA Telescopes; (d) NASA/CXC/SAO]

radiation. The radiated energy is thus taken from the rotation, by which the frequency keeps decreasing over cosmic time scales (10^6–10^8 years) until the rotation eventually is too slow and the synchrotron radiation stops. The initially radiated power is enormous and can reach an equivalent of about 10,000 Suns for several thousand years. From the frequency f and its change over time $\dot{f} = \frac{df}{dt}$, the approximate age of the pulsar can be determined using basic classical equations ($\tau = 2 \cdot f / \dot{f}$). This simple formula can be compared with the time of the supernova event from which the pulsar originated, by either calculating this event back from observational data of the supernova remnant or, in a few cases, by consulting historical records. For example, in the present year 2021, i.e., at the time of writing this book, the age of the Crab Pulsar 6200 light-years away is calculated to about 1350 years (Fig. 11.2). The actual supernova event occurred according to the historical records in 1054, i.e., 966 years ago, which compares reasonably well with the calculated number.

Rotation and Relativity

Of course, the rotational frequency cannot become arbitrarily high, because at about 2380 Hz the surface at the equator of such a star (20 km in radius) would move in a simple classical view with the speed of light!! Relativity will have to prevent this and it does. However, a limit will already be reached earlier, because at a frequency of about 1500–2000 Hz the enormous centrifugal forces in the equatorial region produce instabilities and local warping due to a "softening" of the nuclear compression. Such warping destroys the rotational symmetry and immediately leads to the emission of gravitational waves and thus to an energy outflow that pushes the rotational frequency back to its limit. The power radiated by gravitational waves is proportional to the 6th power (!!) of the frequency ($P \propto f^6$). Gravitational radiation is thus in the end the determining factor for the upper limit of the rotational frequency of a neutron star. If sub-millisecond neutron stars were found in the near future, the theory of neutron stars as well as the theory of relativity could have a problem.

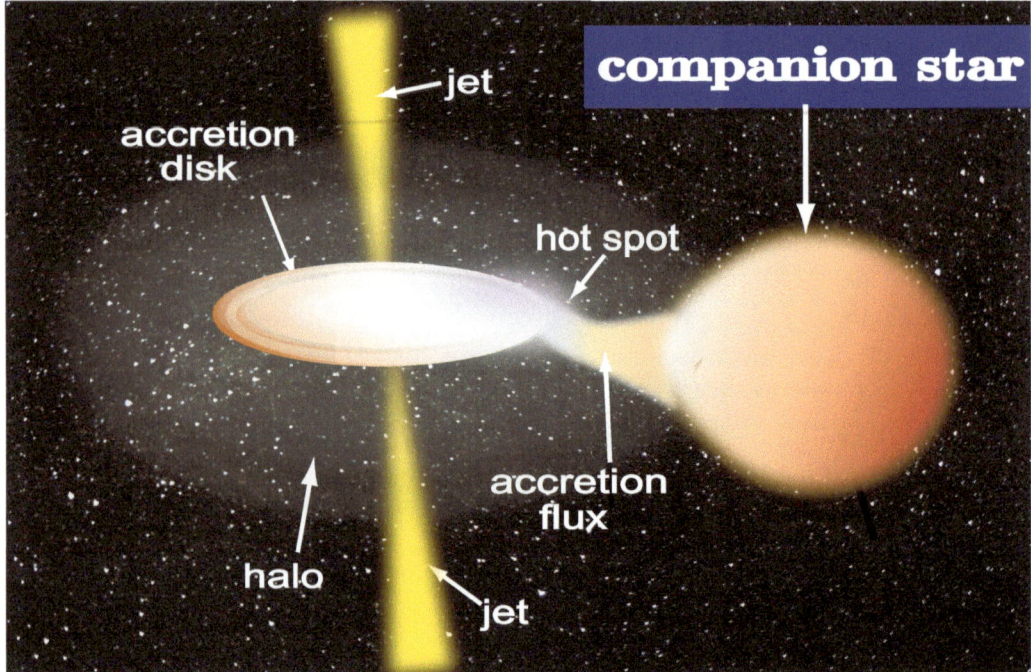

Fig. 11.3: Shown is a graphical representation of a neutron star, which extracts mass from its companion star to convert to a Black Hole. The figure refers to the "SS433" system discovered in 1976 located about 18,000 light-years away. The accreted mass is gravitationally accelerated to about 30% of the speed of light and emits intense X-ray and γ-radiation near the "hot spot" and inside the disk. Note that the two objects are only about 39 million kilometer (equivalent to about 56 solar radii) apart and rotate around each other with a period of 13.1 days. The mass of the companion star has been determined to be (12.3 ± 3.3) and the mass of the neutron star to be (4.3 ± 0.8) solar masses, so it is possible that the neutron star has already converted to a Black Hole.

Collapse to a Black Hole

Neutron stars can increase their rotational frequency due to the accretion of mass from a companion star (e.g., in the case of a binary system) and at the same time complete the transition to a rotating Black Hole. In principle, there is even a chance of witnessing such a transition live. The binary star system "SS433", about 18,000 light-years away, contains such a possible candidate (see Fig. 11.3). The neutron star, which in this figure lies within the accretion disk, was formed from a supernova at some time

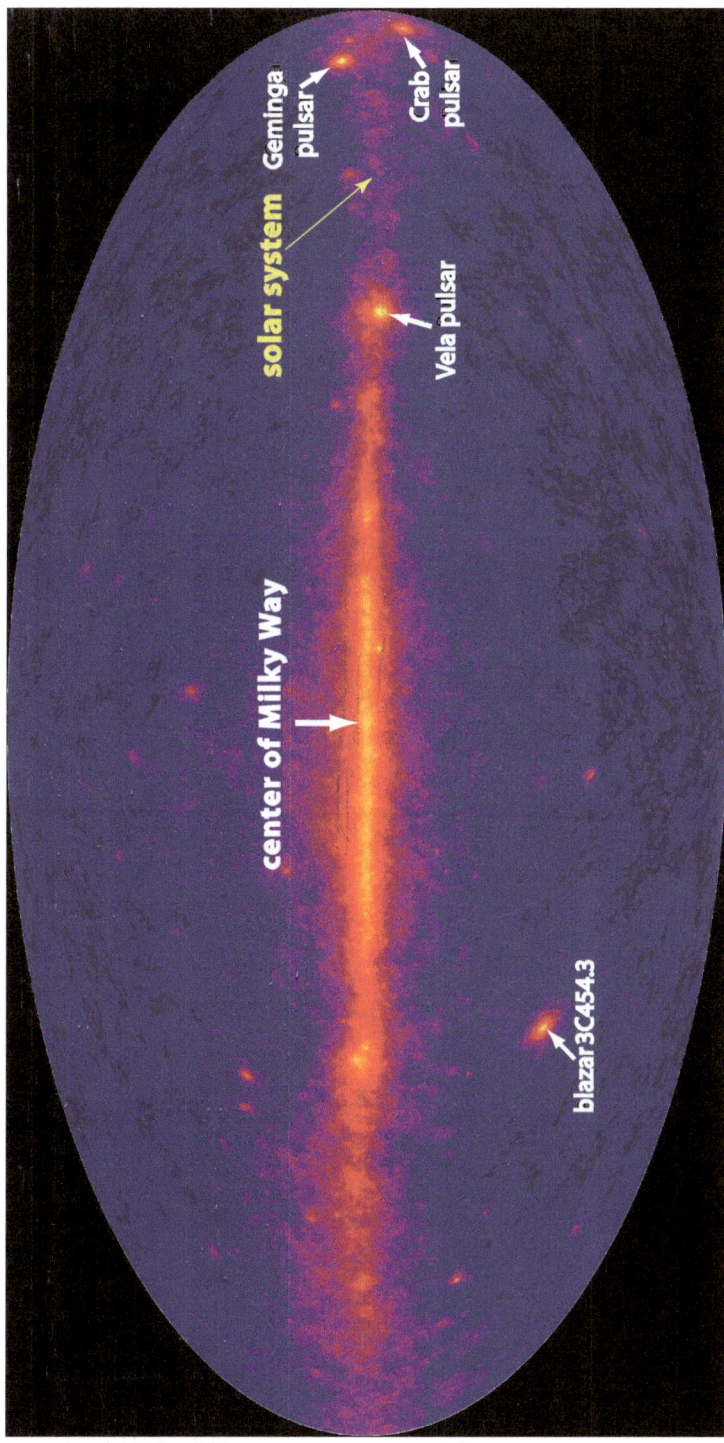

Fig. 11.4: A projection of the Milky Way showing the approximate positions of the two pulsars closest to the solar system, the Vela and Geminga pulsar. Also in the field of view is the projection of blazar 3C454.3, about 7.7 billion light-years away, whose radiative power in periods of high activity is comparable to about 550 billion Suns (its jet emissions are precisely aimed at the Milky Way). It is considered the brightest astronomical object ever observed. Note: The horizontal scale is not linear in this plot. [Public Domain]

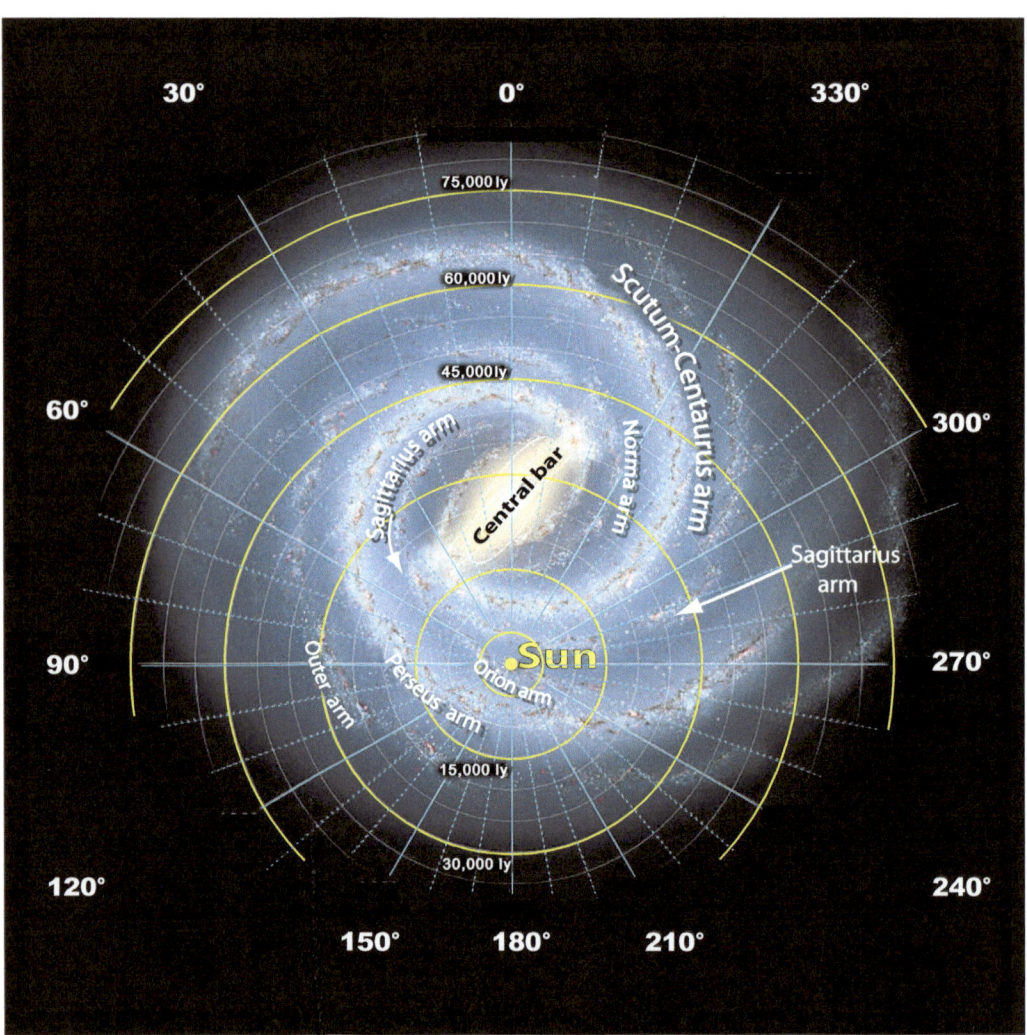

Fig. 11.5: Scaled view of the Milky Way with its different arms. The Sun is inside the yellow disk, which in this representation has a radius of about 1000 light-years. The fact that several neutron stars are found in this range shows that neutron stars are by no means rare objects in a galaxy.

[Credit: NASA/Adler/U. Chicago/Wesleyan/JPL-Caltech, adapted for this book]

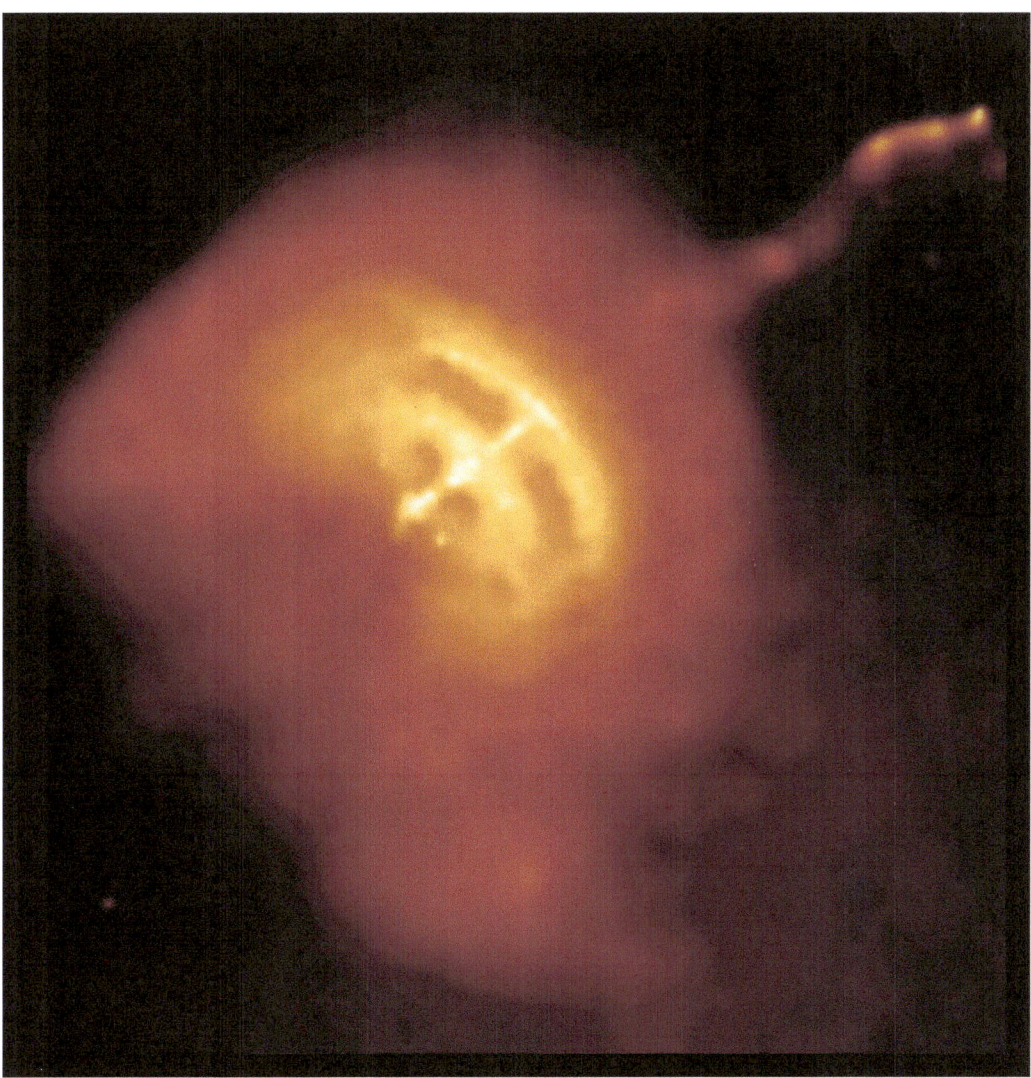

Fig. 11.6: The Vela pulsar with its distinctive pulsar jet, which literally drills a hole into the surrounding nebula and excites it to light emissions. The Vela pulsar rotates at about 11.2 Hz. The length of the jet is about 0.7 light-years and has the shape of a rotating helix, which results from its precession about the rotation axis of the pulsar. In the single frame from a movie sequence shown here, this precessional movement is not immediately visible. [Public Domain]

in the past. It is gravitationally locked to a massive star and extracts from it large amounts of mass. Whether it has already made the transition to a Black Hole has not yet been answered conclusively, though the extremely high-energy radiation it emits with the jets is rather unusual for an ordinary neutron star. These energies reach up to $10^{15} - 10^{16}$ eV [2].

Anatomy of neutron stars

Neutron stars are not directly visible, but can be recognized by their special radiation behavior. This even holds for the two known and probably closest pulsars to the solar system, the Vela and the Geminga pulsars, whose approximate positions within the Milky Way are given in Fig. 11.4 (distance Vela: 935 ± 60 light-years, distance Geminga: 810^{+400}_{-200} light-years, compare also Fig. 11.5). Nevertheless, the structure of a neutron star can be inferred with some certainty from the characteristic light emissions and the emission shapes (see, e.g., Fig. 11.6), so that theoretical models become more and more testable over time as observations become more and more detailed.

In the Fig. 11.7 the anatomy of a neutron star is shown as it is imagined today. Accordingly, a neutron star has an ultra-thin hydrogen "atmosphere" at its surface, compressed to a density of $\rho \approx 10^4$ g/cm^3. This layer is probably only a few millimeters to centimeters thick. This is followed by about a 100 m thick layer consisting of elements in the iron/nickel region. The iron/nickel region is still a remnant from the supernova explosion, from which, after all, the neutron star had emerged at some point. This layer has typical densities of up to a few 10^{11} g/cm^3. If one drills deeper into the star, one reaches a region, which initially still has a small fraction of extremely neutron-rich nuclei, but eventually consists almost entirely of a neutron fluid with a small percentage of free protons. The making of nucleus-like states is no longer possible. Here, densities of about 10^{12} to 10^{13} g/cm^3 are reached, depending on the depth. Beyond 10^{13} g/cm^3 and at a depth of about 5 kilometers onwards one reaches the densest possible packing of neutrons, in

---

[2] T. C. Hillwig, et al., *Identification of the Mass Donor Star's Spectrum in SS 433*, The Astrophysical Journal 615, 422 (2004) or same authors, arXiv:astro-ph/0403634v1, and T. C. Hillwig, D. R. Gies, *Spectroscopic Observations of the Mass Donor Star in SS 433*, The Astrophysical Journal, 676:L37 (2008).

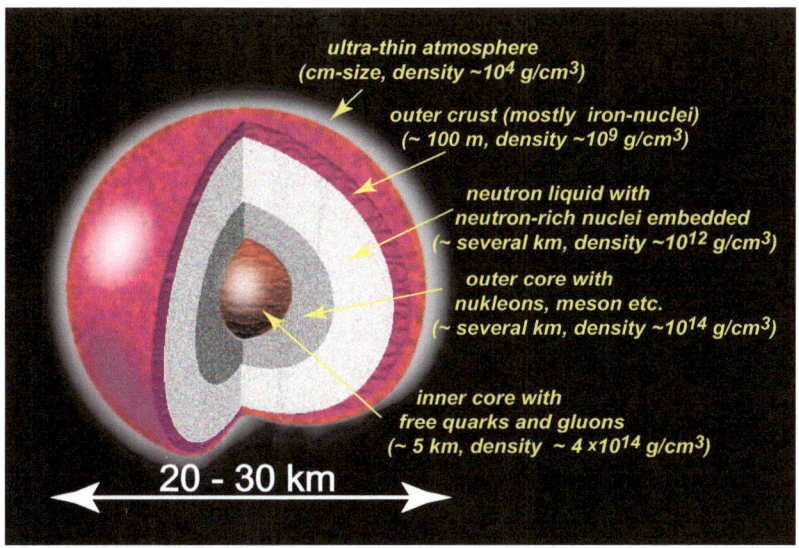

Fig. 11.7: Anatomy of a neutron star.

such a way that they still remain halfway identifiable as neutron objects. In the central region and from about a density of some 10^{14} g/cm^3 (= some 100 million tons per cm^3) the neutrons come so close to each other that they dissolve into their constituents, namely quarks and gluons.

Temperature

The surface temperature of a neutron star lies typically in the range of 10 and 500 Mega-K, and quite curiously, the neutron star cools down only very slowly (which is partly due to its small surface). Nevertheless, the cooling process of a neutron star is still not generally understood. To emit electromagnetic radiation, electric charges need to be moved and accelerated, yet their number is considerably reduced in a neutron star. Pulsars take the energy of their radiation mainly from their rotational energy and not from the temperature bath. Electromagnetic radiation created at a depth greater than a few 100 meters has no chance of ever reaching the surface of the star.

So the question remains: What process provides cooling, and how long does such a cooling process take?

Answer: Next section.

URCA cooling — a "money's gone" anecdote

Urca is a small city near Rio de Janeiro in Brazil, about 3 km from Copacabana. This small place housed a casino, the Cassino da Urca. George Gamow visited Rio de Janeiro in 1940, and he there met Mário Schenberg, a Brazilian astrophysicist, who, as a result of this meeting, joined Gamow later as a Guggenheim fellow at George Washington University in Washington, DC. Still in Urca, both visited the Cassino da Urca, and within a short time they gambled away what must have been a considerable amount of money. Back in Washington, both studied these elusive processes in supernovae, which generate this enormous outflow of energy (most of it invisible) and which happen over fantastically short times. It should be noted here that at that time the physics of supernovae was still largely in the dark. In this context, Schenberg jokingly remarked to Gamow [3]:

"the energy in a supernova disappears as fast as the money at the roulette table in Urca"

George Gamow subsequently adopted the term «**URCA**», especially since this term also had the meaning of "thief" in his native Ukrainian language. He used it as a synonym for the "thieving" neutrinos, which caused the unobservable energy outflow of a supernova. Note that neutrinos were not experimentally verified at that time. The term nevertheless survived and was used after the 1967 discovery of neutron stars and pulsars to refer to the neutrino cooling of neutron stars.

There are two variants of the URCA process, the "direct" and the "modified" URCA process.

The direct process:

$$n \longrightarrow p + e^- + \bar{\nu} \quad \text{immediately followed by} \quad p + e^- \longrightarrow n + \nu$$

produces an anti-neutrino and a neutrino, both of them leaving the star and withdrawing energy from it. This process is extremely efficient, but for thermodynamic and quantum mechanical reasons it is only allowed beyond extreme temperatures ($\gtrsim 500$ Mega-K) and only where the proton fraction in the star is at least 10%.

---

[3] George Gamow, *My World Line — An Informal Autobiography*, The Viking Press, New York, 1970.

The modified process:

$n + n \longrightarrow n + p + e^- + \overline{\nu}$ immediately followed by $n + p + e^- \longrightarrow n + n + \nu$

looks similar and also produces two neutrinos leaving the star, but is several orders of magnitude less efficient and not subject to the above restrictions. The extra neutron is required to maintain a balanced momentum in the densely packed neutron star. We will not go into this detail, which is a quantum mechanical effect for fermions. Nevertheless, this reaction is the determining process for the cooling of a neutron star even below a temperature of about 500 Mega-K, and therefore is also the defining factor for the time scale of its complete cooling, which eventually leads to its invisibility. For comparison, this time scale could be many orders of magnitude beyond the current age of the Universe. Whatever happens to the neutron star in the end remains unclear — the possible, though not yet discovered nucleon decay, would give this object perhaps another 10^{45} seconds or so, until it has finally dissolved into photons and neutrinos (see also page 51).

Dragging of the inertial frame

General Relativity makes an astonishing and also exceedingly curious prediction for rotating systems: A rotating mass has its own co-rotating or co-dragged inertial frame. One may remember the reflections of Ernst Mach about the origin of the so-called "apparent" or "inertial" forces, to which also the centrifugal force belongs. He assumed that the masses of the distant stars, which counter-rotate to a rotating object,

Fig. 11.8: Resting and relaxing in your own inertial system.

create an "apparent" centrifugal force. In the interstellar (or intergalactic) space this counter-rotation of the distant stars is the only possibility to determine one's own rotation and thus the direction of the centrifugal force.

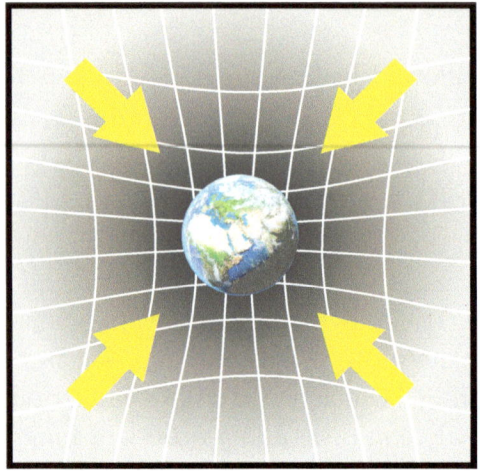 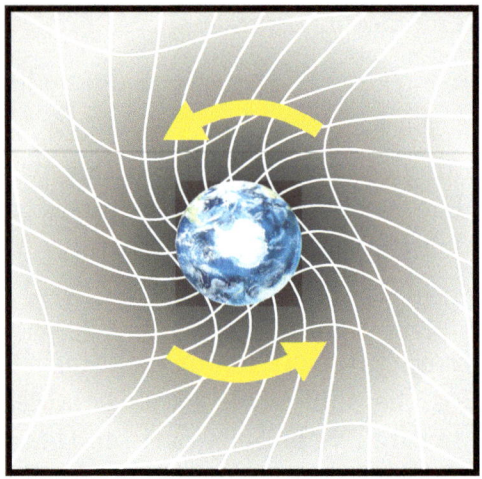

Fig. 11.9: Left: A stationary inertial system. Objects fall in a straight line towards the center of the Earth. **Right:** The "dragging" of the inertial system. Objects follow the inertial system when falling towards the center of the Earth

If Ernst Mach's conjecture holds, then this effect should also come into play for a nearby mass, e.g., that of a neutron star. In fact, General Relativity makes just such a prediction. The centrifugal forces of a rotating neutron star are partially compensated by the dragging-along of the inertial frame. Depending on the internal structure of the neutron star, which is, of course, largely model dependent, the dragging of the inertial frame can account for up to 40% of the rotation. An observer at the equator of the star notices this only to a limited extent. The starry sky still rotates in the opposite direction, but the observer feels only 60% of the centrifugal force and assumes his rotation is less fast.

In the limiting case a curiosity occurs: The starry sky rotates with the rotational frequency of the neutron star (e.g., up to 1000 times per second, depending on size), without the observer feeling a centrifugal force, or even knowing that he is rotating at all. His force-free inertial frame rotates with him (Fig. 11.8). But beware: He now feels the full attractive gravitational force. However, this extreme case will not occur in reality.

Earth, as well, is a nearby mass, and the question arises as to what extent this effect plays a role in the rotation of the Earth (see Fig. 11.9) and how this could be captured by a measurement. We come back again to the

Foucault pendulum under which the Earth rotates. According to General Relativity, the inertial system carried by the Earth would have to drag the rotating pendulum in a way to slightly reduce the apparent 24-hour rotation at the poles. — Memories of the geocentric world model are awakened, which was the common narrative until the 18th century:

Question: *Is the Earth perhaps the center of the Universe after all, around which all stars including planets and the Sun revolve?*

Or: *Does the geocentric worldview now re-appear through the back door?*

Answer: *Yes, but only a tiny bit.*

In the case of the Foucault pendulum, there is an entire 360° rotation missing after about 30 million years, or synonymously, after about 30 million years, an additional sidereal day has accumulated, without a real rotation. At the equator, the centrifugal force is reduced by a factor of 6.1×10^{-12} compared to a purely classical, Newtonian theory, and a human who brings 100 kg on the scales weighs about 0.063 micrograms more due to this missing centrifugal component. The 100 kg human will not notice this, he lacks the comparison with the "classical, Newtonian weight".

11.2   The "Gravity-Probe-B" experiment

The dragging of the inertial frame by a rotating mass is at first only a prediction by General Relativity. Because of the comparatively tiny mass of the Earth, it was long considered immeasurable[4]. Proposals to install a Foucault pendulum at the South Pole and track its rotation over the course of a year with a daily accuracy of 0.01 milli-arcsecond ($\approx 3 \times 10^{-9}$ angular degrees) ultimately proved unrealistic. In order to chart the pendulum's changing distance to the geographic South Pole on the ice of about 10 m/yr would have required a continuous tracking with an approximately 10 cm

---

[4] This effect of General Relativity is also called the "Lense-Thirring" effect after the two Austrian physicists Josef Lense (1890–1985) and Hans Thirring (1888–1976), who first formulated this effect in 1918 in the context of General Relativity. The name Hans Thirring appears again in a different context on page 273.

accuracy. In addition, there would have been further interfering effects to be accounted for with extreme accuracy, e.g., those caused by the Earth's magnetic field, by cosmic radiation, by the minimal fluctuations of temperature and pressure in a specially constructed vacuum container, and also those caused by friction effects of numerous origins, which would not always be controllable. A summary is given, for example, in reference[5].

Of course, the question arises of who should benefit from this tiny effect, even if it were experimentally proven. In this context one may remember the tiny relativistic effects that gave rise to the deflection of light in the gravitational field of the Sun and which initiated the departure from classical Newtonian theory. This rather innocuous observation set in motion a technical development, which in a mere 100 years changed the quality of modern life lastingly, and this in an unforeseeable and positive way (e.g., internet, cell phones, time standards, laser and computer technology, ordinary navigation and satellite navigation, GPS, etc . . .). Today's modern life would be unthinkable without taking into account General Relativity.

As a further reminder: It was already mentioned that General Relativity is incomplete and has to be replaced by an even more comprehensive theory. How this theory should look, is unknown. There is only the hope that it will be once again the small discrepancies to a hitherto common theory that will provide the starting points for new physics and new technologies.

The "Gravity-Probe-B" experiment is to be understood quite in this sense. It is a mission initiated by Stanford University and NASA in order to subject two predictions of Einstein's General Relativity to a decisive test in a way that has never been done before, as there are:

① **The geodesic space-time effect** — This is equivalent to the space-curvature effect as a consequence of the Earth's mass. If one draws a circle around the Earth and takes as a basis for it a Euclidean geometry, then about $3 - 4\,\mathrm{cm}$ would be missing at the circumference of this circle (also called "the missing inch").

So much the theory — the question is: is that so?

---

[5] V. B. Braginsky, A. G. Polnarev, K. S. Thorne, *Foucault Pendulum at the South Pole: Proposal For an Experiment to Detect the Earth's General Relativistic Gravitomagnetic Field*, Physical Review Letters 53, 863 (1984).

(2) **The frame-dragging of the inertial systems** — An inertial system is like a viscous fluid, within which the Earth rotates. Pulled along and twisted by the Earth's rotation it will complete a full 360° rotation at the Earth's surface in 30 million years. From the view of an inhabitant of Earth, this appears as if a whole rotation were missing in 30 million years. On the other hand, a distant observer sees the Foucault pendulum making one revolution around itself instead of remaining stationary and aligned with him/her (Fig.11.10).

So much for theory — again, the question is: is that so?

Of course, NASA had a vested interest in this experiment. The acquired knowledge has a direct impact on the precision of the Global Positioning System (GPS). Furthermore, it is directly used for satellite navigation during interplanetary missions and even more so for the early and precise determination of the paths of massive meteorites, which, coming from outside the inner planetary system, are on collision course with Earth. The slightest perturbations caused primarily by the massive outer planets are significant for course determination.

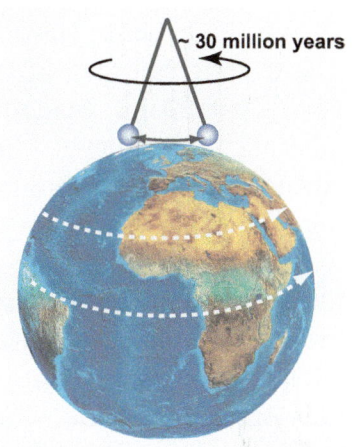

Fig. 11.10: The pendulum and General Relativity.

The basic idea of the "Gravity-Probe-B" experiment is identical to that of the Foucault pendulum. Only the "back-and-forth"-movement of the pendulum is replaced by a gyroscope, whose rotation axis is always perpendicular to the rotation axis of the Earth, as it is in the case of the pendulum. The observable quantity (from the point of view of a distant observer) is then the rotation of the gyro axis in a completely force-free inertial frame — or in the local space-time — of the rotating Earth. This sounds simple at first, but the technical problems involved required nearly 60 years of continuous development.

Force-free translates to weightless, and that in turn implies that the experiment can only be conducted in a satellite orbiting the Earth in as low an orbit as possible.

Here are some of the most important and at the same time most remarkable parameters of this project:

► **The gyros**

At the heart of the gyroscopic instrument were four high-purity fused silica spheres, 3.8 cm in radius, manufactured to perfection. A maximum deviation from the spherical shape of 55 nm had been specified for these spheres, and a value of less than 33 nm was achieved, making the spheres the most accurate spherical objects ever made. Quartz glass was used because, on the one hand, it is extremely homogeneous due to its amorphous structure

Fig. 11.11: The gyroscopes of the "Gravity-Probe-B" experiment during fabrication. **Left:** raw mold made of quartz. **Right:** final mold with the niobium coating.

(relative density deviation $\Delta\rho/\rho \leq 3 \times 10^{-7}$), and on the other hand, it is one of the materials with the lowest thermal expansion coefficient.

The spheres were coated with an ultra thin layer of the superconducting metal niobium (transition temperature 9.25 K) (see Fig. 11.11) and finally cooled in the experiment to the temperature of liquid helium (1.8 K) and then centered in a special holder, where they were kept contact-free by virtue of an electric field. In addition, magnetic shielding existed against the Earth's magnetic field.

► **In orbit**

On April 20, 2004, a Delta-II rocket finally brought the fully equipped gyroscopic assembly with the four gyros into a 642 km high, polar and nearly perfect circular orbit (deviation from circularity about 12 m). The time for one revolution was 97 minutes and 36 seconds.

Once in the target orbit, the gyros were free-floating inside the satellite and aligned with the star "IM Pegasi" (synonymous with the "distant

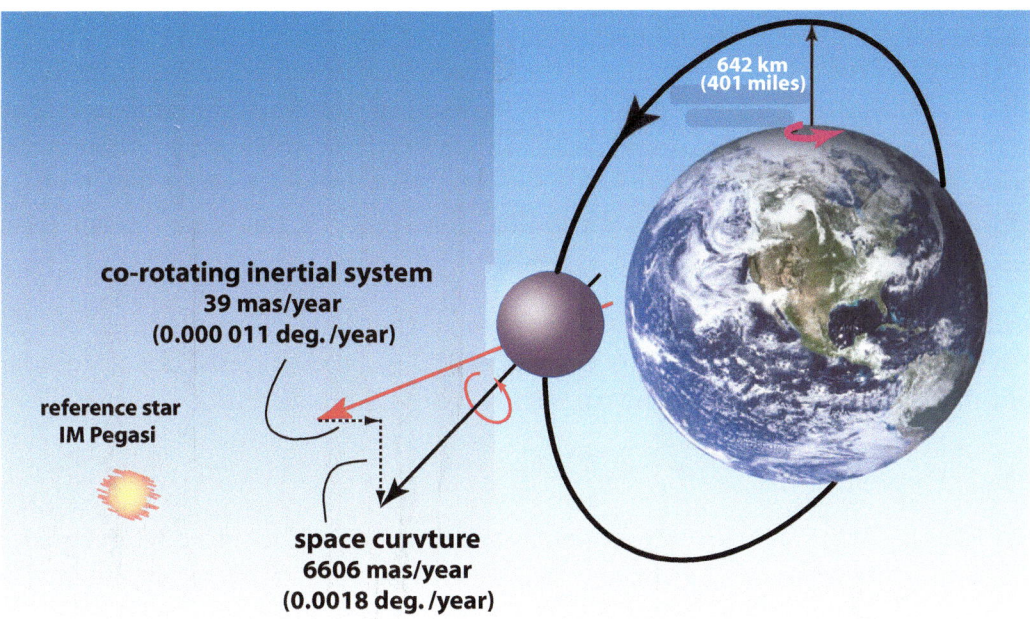

642 km
(401 miles)

co-rotating inertial system
39 mas/year
(0.000 011 deg. /year)

reference star
IM Pegasi

space curvture
6606 mas/year
(0.0018 deg. /year)

Fig. 11.12: Representation of the gyroscopic motion in the "Gravity-Probe-B" experiment. The experiment is aligned with the star "IM Pegasi", 300 light-years away, whose position and proper motion are known with extreme precision. The change in gyro-orientation is predicted by General Relativity to be 6606 mas (milli-arcseconds) in the north-south direction and 39.2 mas in the west-east direction per year. The measured values are 6601.8 ± 18.3 mas and 37.2 ± 7.2 mas.

observer"), which is 300 light-years away, and whose position and proper motion are particularly well known (see Fig. 11.12). In order to compensate gravitational forces originating from the surrounding satellite mass, the satellite was put into a 77.5 second self-rotation.

▶ Compensation of forces acting on the satellite

In the relatively low orbit, decelerating forces caused by the Earth's outer atmosphere continue to act. Without continuous compensation, a satellite would steadily lose speed and altitude. Gas exhaust valves therefore regulated the necessary thrust and thus kept the satellite and gyroscope in position.

However, non-compensatable gravitational forces and torques do remain. These are caused by the inhomogeneous mass distribution of the Earth and give rise to lurching movements along the satellite's orbit. In the "Gravity-Probe-B" experiment, these forces were specified as less than $10^{-11} \cdot g$ ($g = 9.81\,\mathrm{m/s^2}$, or N/kg, i.e., acceleration at the Earth's surface). This corresponds to a gravitational force, which two $1\,\mathrm{kg}$ spheres at a distance of about $0.8\,\mathrm{m}$ exert on each other, or — in a somewhat amusing form — to a force that brings the leg of a mosquito to a speed of $1\,\mathrm{m/s}$ ($= 3.6\,\mathrm{km/h}$) in about $30\,\mathrm{min}$.

The success of the experiment demanded that all! forces acting on the gyroscope be limited to a maximum of $10^{-10} \cdot g$, and, in fact, one succeeded in undercutting this value by an entire order of magnitude.

▶ **Experiment**

With the start of the experiment, the four gyros inside the gyroscope assembly were made to rotate, two clockwise and two counterclockwise. They were then left to themselves at a rotational frequency of $80\,\mathrm{Hz}$. The time constant of rotation, i.e., the time until the gyros were to lose about 37% of their initial rotational frequency, had been determined in advance to be about $15,000$ years !!

At an ambient temperature of the gyroscopes of $1.8\,\mathrm{K}$ (the temperature of liquid helium), the niobium coating mentioned above is superconducting. Now a curious property of superconductors comes into play — rotating superconductors produce a magnetic field (or a magnetic dipole moment) independent of the material, whose strength is proportional only to the frequency of rotation. The property is known as the London moment and is named after the German-American physicist Fritz London (1900 – 1954), who first correctly interpreted this effect[6]. In this particular case, the

---

[6] The phenomenon of magnetic field generation in rotating superconductors was already predicted as early as 1933 by R. Becker, F. Heller and F. Sauter [Zeitschrift f. Physik 85, 772 (1933)]. However, their theoretical derivation was based on some inherently incorrect assumptions. Fritz London pointed this out in 1950 [*Superfluids*, John Wiley & Sons, New York, Vol. 1, Sec. B, (1950)] and provided a correct explanation, which is why the effect bears his name. The later developed BCS-theory of superconductivity by Bardeen, Cooper and Schrieffer also describes this phenomenon, though in an entirely incorrect way up to a wrong direction of the magnetic field. It was not until 1964 that the London moment could finally be determined experimentally by A. F. Hildebrandt [Physical Review Letters 12, 190 (1964)], in agreement with the theoretical prediction made by F. London.

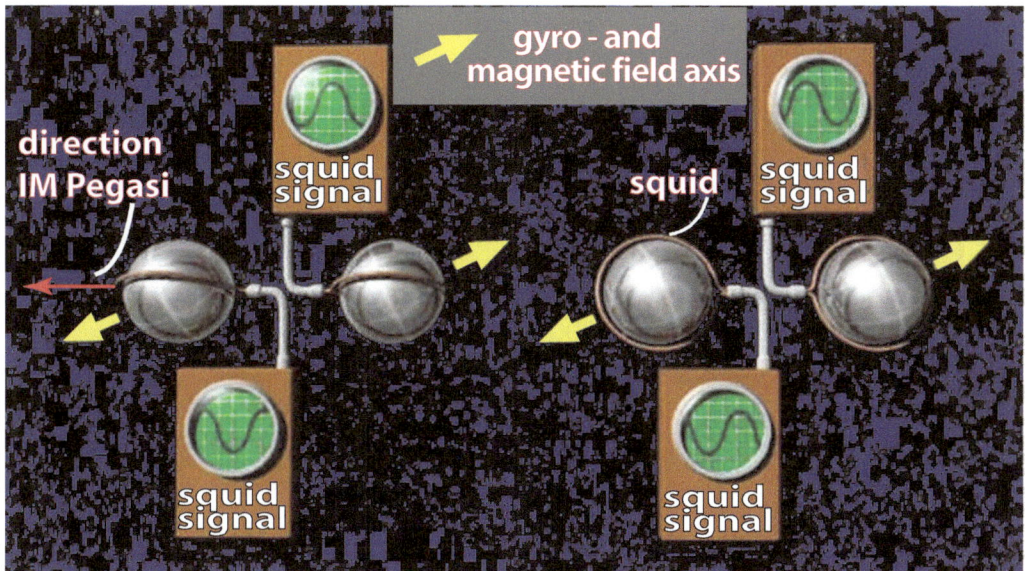

Fig. 11.13: Determination of the gyro-axis in the "Gravity-Probe-B" experiment using a squid magnetometer. The squid rotates around the gyros and records a phase signal, from which the orientation of the magnetic field lines penetrating the squid loop can be determined with highest precision.

direction of the magnetic field axis is identical to the direction of the rotation axis.

The magnetic field of a rotating superconductor that is generated at a frequency of 80 Hz has a strength of 5.7×10^{-9} Tesla. In comparison: The Earth's magnetic field has a strength of about 70×10^{-6} Tesla at the poles and about 25×10^{-6} Tesla at the equator. With appropriate magnetic shielding, the residual field strength in the "Gravity-Probe-B" experiment could be lowered to below 10^{-11} Tesla, which was more than two orders of magnitude smaller than that of the rotating superconductors.

The determination of the magnetic field properties in the "Gravity-Probe" gyroscope was accomplished by using "squids" (squid = superconducting quantum interference device). Squids are extremely high-resolution magnetometers and capable of detecting minute magnetic field strengths even below 10^{-18} Tesla. In the gyroscope assembly, they revolved around the gyros, so that both, axis position and strength of the field enclosed by the squid loop, could be determined from the signal.

A schematic view of the gyroscope with the squid magnetometers in a fixed alignment with the star IM-Pegasi is shown in Fig. 11.13.

▶ **An astonishing result**

The "Gravity-Probe-B" experiment was carried out over a period of about 17 months until the supply of liquid helium was finally used up. The actual measurement period lasted about 12 months, and during that time the satellite completed about 5000 Earth orbits. Unexpected noise effects discovered only during the mission complicated the analysis, so that the final results could not be published until 2011, and unfortunately they also had a larger error than initially specified [7]. More than 50 years had passed from the beginning of the project in 1959/60 until its end, and more than 500 young scientists (i.e., PhD students) contributed to this experiment in one way or another.

As a result, with this experiment the theory of General Relativity has passed one of its most important tests with flying colors — although one would almost be inclined to say, "unfortunately", because a deviation from General Relativity would have been considerably more exciting. In detail:

● at the altitude of the satellite, the inertial system of the Earth was found to co-rotate with the Earth in west-east direction by:

Gravity-Probe-B measured:	**37.2±7.2 mas/year**
General Relativity:	**39.2 mas/year**

Remember: 39 mas corresponds to the width of a human hair (0.1 mm) observed from a distance of about 530 meters.

● at the altitude of the satellite, the space curvature of the Earth was found to cause a rotation of the gyros in north-south direction by:

---

[7] C. W. F. Everitt et al., *Gravity Probe B: Final Results of a Space Experiment to Test General Relativity*, Physical Review Letters 106, 221101 (2011).

Gravity-Probe-B measured: **6601.8±18.3 mas/year**
General Relativity: **6606.1 mas/year**

This implies that, by assuming a flat Euclidean geometry, about 4.2 cm are missing for each circular orbit of the satellite. With respect to a circumferential circle near the Earth's surface, this comes to 3.8 cm (1.5 inch ⇒ the missing inch).

1 mas=1 milli-arcsecond $= 2.778 \times 10^{-7}$ degree
$\qquad\qquad\qquad\qquad\quad = 4.848 \times 10^{-9}$ radian [or meter/meter]

11.3   Black Holes

Black Holes are probably the most bizarre and the least understood objects in the Universe. Moreover, they are by no means rare phenomena. Every giant galaxy — as far as it is known — hosts a supermassive Black Hole in its center, whose mass can be up to 10 billion solar masses. In comparison, the Black Hole in our own Milky Way Galaxy, with its approximately 4.2 million solar masses, is comparatively tiny. In addition, about 10% of all supernova explosions in a galaxy leave behind a Black Hole. This means that in our Milky Way Galaxy there could well be about $20-40$ million Black Holes from past supernova explosions.

Also the "Big Bang" could have produced mini-Black Holes 13.8 billion years ago, which still vagabond in the extragalactic environment and where they may decay explosively every now and then. However, no one has yet observed such an event. We will also see on the next pages that such mini-versions of Black Holes would have to have a minimum mass of about 10^7 tons to be able to survive to the present time.

All cosmic (!!) Black Holes have a cosmic voracity in common — whoever comes too close to them is inevitably and irrevocably swallowed up and begins a journey into an unknown cosmic Beyond. The "excretion" product of Black Holes is believed to be only radiation, and if the radiated energy is not compensated by a steady increase of mass, a Black Hole (as mentioned above and at the beginning of this book) will evaporate in a cosmic mini-explosion at the end of its existence. This is generally thought to be the case when applying current theories.

Yet, for a start it may be useful to get a clearer picture about what Black Holes are, and to examine the scenarios that develop if the known laws of physics are extrapolated towards extreme masses and extreme gravitational potentials. To what extent this also corresponds to reality in the end will have to be left open for the time being.

Schwarzschild metric, Schwarzschild radius

In the presence of a mass the space-time structure changes; this was already described at different places in this book and also observed experimentally. But how do the changes look like, if the masses become so extreme that even light remains trapped, although the speed of light is an invariable constant?

Shortly after the publication of Einstein's field equations, Karl Schwarzschild $(1873-1916)$ had already solved these equations for a number of simplifying assumptions and was able to derive a metric for space-time. The metric is named after him as Schwarzschild metric and describes with the following equation already the essential and for this book relevant physics concepts (for the sake of keeping it simple, the angle-dependent terms are omitted):

$$\Delta\tau^2 = \left(1 - \tfrac{2GM}{c^2 r}\right)\Delta t^2 - \tfrac{1}{c^2}\left(\tfrac{1}{1-\frac{2GM}{c^2 r}}\right)\Delta r^2$$

$$= \left(1 - R_s/r\right)\Delta t^2 - \tfrac{1}{c^2}\left(\tfrac{1}{1-R_s/r}\right)\Delta r^2$$

with

$$\tfrac{GM}{r} \quad \Longleftarrow \quad \textbf{gravitational potential}$$

$$R_S = \tfrac{2GM}{c^2} \quad \Longleftarrow \quad \textbf{Schwarzschild radius}$$

The metric indicates how the space-time coordinate $\Delta\tau^2$ of an object near a mass transforms to the space and time coordinates Δt^2 and Δr^2 of a distant observer.

A simple example may clarify this:

If an object is inside and an observer outside a gravitational potential, and both remain stationary at a fixed location at distance r_0, then $\Delta r^2 = 0$. In this case, a time interval $\Delta\tau$ in the object system appears shortened to the distant observer's system by the amount $\sqrt{1 - \frac{2GM}{c^2 r_0}}$, or the observer feels that he is ageing faster with respect to the object — or synonymously, his clock runs faster than the one of the object. We had already encountered this phenomenon in the clock experiment of Hafele and Keating, and also in the phrase *"on a mountain one ages faster than on the coast"* — see page 192 and following.

If the object in the above formula approaches the Schwarzschild radius R_S, the expression in the first bracket becomes "0" (zero) and in the second "∞" (infinity). The object has "gone away" from the observer — even a light signal does not reach the observer anymore. The object now seems infinitely far away and is on the boundary of the Black Hole. A return from this edge is not possible because this would mean that there is a time after an infinite time.

In Fig. 11.14 the situation for an object B falling into a Black Hole and observed by A (and in the reference system of A) is again clarified. B continuously emits light pulses, which are received by A at increasingly longer time intervals (Δt_1, Δt_2, ...). The light emission cone of B contracts more and more, this according to the above equation. Meanwhile, the time in A continues "normally", while in B it seems to run slower and slower, until finally the signal emitted at the Schwarzschild radius R_S (or at the event horizon) does not reach the observer A anymore in a finite time.

For the traveler B to the Black Hole the situation is at first uncritical. He has his own proper time and for him it does not change. However, the closer this traveler gets to the Schwarzschild radius, the more his own head

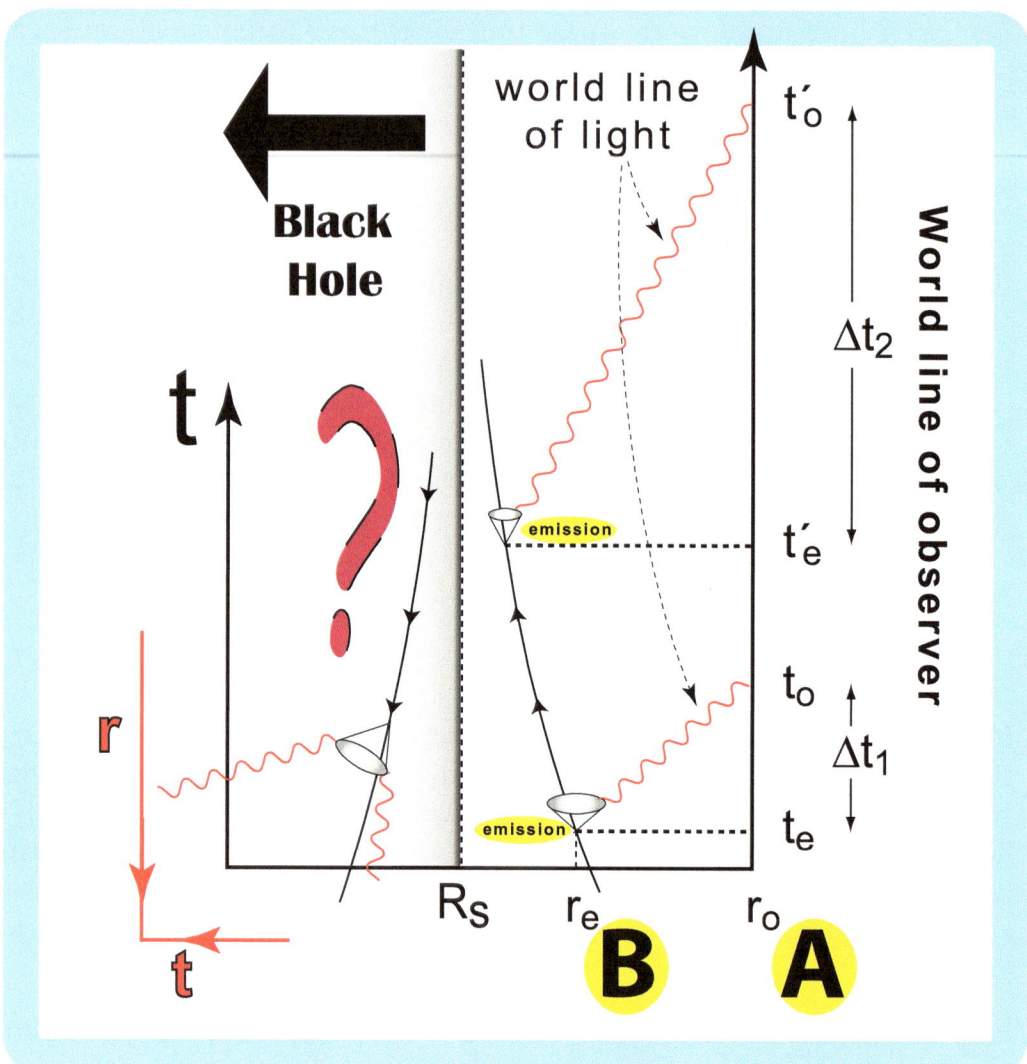

Fig. 11.14: World-lines (location versus time) near a Black Hole. Object B falls into a Black Hole and is observed by A. A is stationary at location r_0. Object B continuously emits light pulses each at emission points r_e, which are received by A at increasingly longer time intervals (δt_1, δt_2, ...). The emission cone of B contracts more and more until finally the signal emitted at the Schwarzschild radius R_S does not reach the observer A in a finite time. A sees time standstill for B. The jump over the singularity (infinity) and the following physics inside a Black Hole is unknown. Indicated here is a special metric in which the space-time parameters of position and time are exchanged — position becomes time and time becomes position (in red). This is mathematically correctly derived according to General Relativity, from the intuition, however, strange.

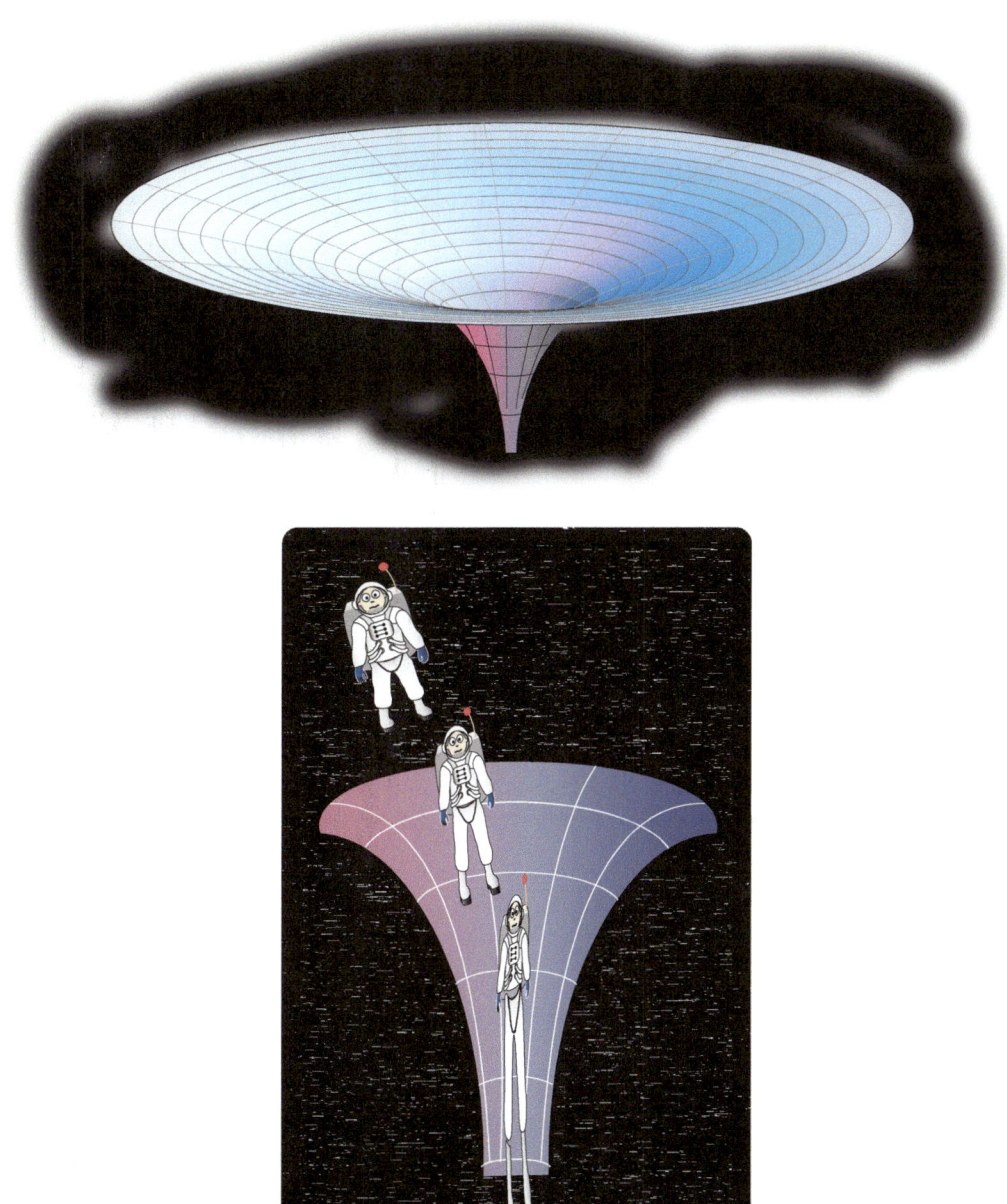

Fig. 11.15: Top: Attempt to represent space-time curvature in 3-dimensional projection.
Bottom: The hypothetical fall into the Black Hole.

becomes an "observer" for his feet, which now move further and further away from the head and eventually disappear into infinity. Moreover, the gravitational differences between head and feet will tear him apart in time. The situation is shown in the Fig. 11.15 in a somewhat simplified way.

Nevertheless, let's assume that the traveler B succeeds in jumping over the singularity unharmed. What awaits him now is unclear and largely speculative. Of course, there are approaches in theoretical physics to connect a metric in the inner region of the Black Hole to the outer Schwarzschild metric, bypassing the physically unreasonable singularity, whereby also quantum mechanical effects or even extra dimensions are brought into play. However, common to all metrics, a stationary situation for B no longer exists, and direction reversal is also excluded. Object B is on an ever accelerating journey to an unknown destination. Observer A is detached. Information from or about B no longer penetrates to the outside.

This is of course an extremely unsatisfactory situation for physics. Therefore, it will remain a central task of experimental astrophysics to find out which information can nevertheless be extracted from Black Holes, in order to be able to advance and test theoretical models. Worm-Holes, White Holes or Parallel Universes, which are often mentioned in science fiction, are clearly outside the current understanding of physics, so, for the time being, they will have to remain what they are: fictions.

Sizes of Black Holes, spaghettification

A common misconception about Black Holes is that they consist of extremely compact matter exceeding in density that of all other objects in the Universe. This view is not correct in this general form. Since the Schwarzschild radius is proportional to the mass M, the density ρ then is:

$$\rho = \frac{M}{V} \propto \frac{M}{R^3} \propto \frac{1}{M^2},$$

i.e., with increasing mass, the density becomes quadratically smaller. This fact is listed for some objects in the following Table 11.2.

One can see that only the comparatively small masses, like the one of the proton, the human being or even like the mass of the Earth, feature extreme and completely unrealistic densities when transformed into Black Holes. And while even the Sun would still have to be compressed to about 40 times the density of a neutron star to sink into a Black Hole, the densities of supermassive Black Holes at the centers of galaxies are comparatively small. For example, a Black Hole of 1 billion Suns (!!) could easily be accommodated in the interior of our solar system. The Schwarzschild radius of such a Black Hole would then be about 2/3 of the distance to the planet Neptune, but the density of $18\,\mathrm{kg/m^3}$ ($=0.018\,\mathrm{g/cm^3}$) would be just about the density of particularly light styrofoam. The same is true for a whole galaxy, whose Schwarzschild radius is about half a light-year. If the mass were homogeneously distributed, the density would be comparable to that of a "moderately good" vacuum on Earth (about 10^{-2} mbar).

An interesting situation arises when one calculates the Schwarzschild radius of the Universe. Its visible mass is about 10^{53} kg, and from this, the Schwarzschild radius is calculated to be 14.8 billion light-years, which is about the size of the visible Universe and therefore also equal to its present density. Coincidence? — maybe, but coincidences are extremely unpopular in physics.

Another interesting and also amusing situation arises when the radius of a Black Hole becomes particularly small, as would be the case, for example, for the Moon ($R_S \simeq 100\,\mu\mathrm{m}$) or the Earth ($R_S \simeq 9\,\mathrm{mm}$). For the astronaut in the Fig. 11.15 this would result in a "spaghettification". Not only would he be stretched out, but the transverse forces would also compress him to below 10^{-10} m (about the diameter of an atom). But be careful: these nonsensical conclusions appear only if one applies the known laws of physics without corrections to these extreme scales.

Tab. 11.2: The table lists characteristic values, such as Schwarzschild radius, density, lifetime and temperature of Black Holes of different sizes. The numbers highlighted in yellow are not particularly plausible according to the current understanding of physics. The theoretical model of Stephen Hawking is likely incomplete here. Note: The more massive a Black Hole, the larger the Schwarzschild radius and the smaller the density. The mass of the Black Hole in our Milky Way Galaxy is comparatively small at 4.2 million solar masses. It has a Schwarzschild radius of about 18 solar radii and a density about 100 times that of lead. In contrast, a 1 billion solar mass Black Hole has a density of only super-light styrofoam ($18\,\mathrm{kg/m^3} = 0.018\,\mathrm{g/cm^3}$).

When inferring Hawking radiation, the temperature of Black Holes with masses from about the mass of the Earth onwards will be smaller than the temperature of the Universe for at least the next 10^{19} seconds (about 300 billion years). Over this time the Black Hole will receive energy from the temperature of the Universe, and only thereafter the situation reverses, when the temperature of the Universe has dropped to about 10^{-2} K.

A mini-Black Hole (mini-BH) with the mass of 170 million tons would have approximately the diameter of a proton and a lifetime equal to the age of the Universe.

Parameters of Black Holes of different sizes

object	mass [kg]	Schwarzschild radius	density [kg/m³]	decay time [s]	temp. [K]
proton	$1.4 \cdot 10^{-28}$	$2 \cdot 10^{-55}$ m	$3.7 \cdot 10^{135}$	$2 \cdot 10^{-100}$	$9 \cdot 10^{50}$
human	70	$1 \cdot 10^{-25}$ m	$1.5 \cdot 10^{76}$	$3 \cdot 10^{-11}$	$2 \cdot 10^{21}$
meteorite	$1 \cdot 10^{6}$	$1.5 \cdot 10^{-21}$ m	$7.3 \cdot 10^{67}$	84	$1 \cdot 10^{17}$
mini-BH	$1.7 \cdot 10^{11}$	$2.5 \cdot 10^{-16}$ m	$2.5 \cdot 10^{57}$	$4.1 \cdot 10^{17}$	$7 \cdot 10^{11}$
moon	$7.3 \cdot 10^{22}$	$110\,\mu\mathrm{m}$	$1.4 \cdot 10^{34}$	$3 \cdot 10^{52}$	1.7
Earth	$6 \cdot 10^{24}$	8.9 mm	$2.0 \cdot 10^{30}$	$2 \cdot 10^{58}$	$2 \cdot 10^{-2}$
Sun	$2 \cdot 10^{30}$	3.0 km	$1.8 \cdot 10^{19}$ $\approx 40 \times \rho_{\mathrm{n\text{-}star}}$	$7 \cdot 10^{74}$	$6 \cdot 10^{-8}$
4.2 million Suns	$8.4 \cdot 10^{36}$	12.4 million km $\approx 18 \times R_{\mathrm{Sun}}$	$1.1 \cdot 10^{6}$	$5 \cdot 10^{94}$	$2 \cdot 10^{-14}$
1 billion Suns	$2 \cdot 10^{39}$	2.6 light-hours $\approx \frac{2}{3} \times$|Sun-Neptune|	18	$7 \cdot 10^{101}$	$6 \cdot 10^{-17}$
galaxy	$\approx 3 \cdot 10^{42}$	0.45 ly	$\approx 8 \times 10^{-6}$	$2 \cdot 10^{111}$	$4 \cdot 10^{-20}$
Universe	$\approx 10^{53}$	≈ 14.8 billion ly	$\approx 7.3 \cdot 10^{-27}$ $\approx \rho_{\mathrm{Universe}}$	$8 \cdot 10^{142}$	$1 \cdot 10^{-30}$

Hawking-Radiation and lifetime

If a Black Hole were "totally black", then its surface temperature would have to be at absolute temperature zero ($T = 0$ K). But since such a Black Hole is colder than the Universe ($T_{\text{Univ.}} = 2.73$ K), thermal energy from the Universe would continuously flow into the Black Hole. Thus, the mass of the Black Hole would continuously increase according to the equivalence of mass and energy, and its size expand, yet the surface temperature would have to remain unchanged at $T = 0$ K. Such behavior is completely incompatible with the laws of thermodynamics, since this would be equivalent to a *«perpetuum mobile»* [8]. Therefore, thermodynamics dictates unmistakably that the surface temperature of a Black Hole has to be finite, a requirement which at the same time shows that the singularity at the event horizon is unreasonable.

For the reverse case, when the surface temperature is greater than the temperature of the Universe, the Black Hole —this time conforming to thermodynamics — will give away energy (i.e., temperature) to the Universe. The Black Hole then has the property of shrinking, or decaying with time. In this case, the question arises: What mechanism gives the Universe a temperature, what value does it have, and what is the decay constant of the Black Hole?

Stephen Hawking was the first, who approached this question in the context of a quantum field theory and derived a formula for the temperature radiation of a Black Hole, this, however, in a semi-classical approximation. The radiation is named in the literature after him as Hawking radiation. Strictly speaking, the formula is only usefully applicable for extremely massive Black

---

[8] A simple explanation: If a Black Hole increases in size, it performs work against the external pressure (e.g., against the radiation pressure) of the Universe. However, thermal energy cannot be completely converted into work, so the Black Hole must return some of the heat back to the Universe. This is the principle of a heat-engine. If this principle could be circumvented, one could construct a *«perpetuum mobile»* e.g., in the form of a vehicle that sucks in ambient air, cools it, uses the released energy for motion (work), and releases the cooled air again to the environment — completely without fuel and completely without external energy supply. Even until today, tinkerers, in ignorance of the laws of physics, try to construct such a machine — without success.

Holes because for such objects the space-time curvatures require only a small correction to the formula. For comparatively small Black Holes (then, when the above quoted spaghettification assumes nonsensical dimensions), a quantum gravity theory is required, which so far does not exist. From such a theory one expects a shortening of the lifetime compared to the semi-classical formula of Hawking.

In Table 11.2 some characteristic parameters, like for instance "decay time" and "temperature", are calculated according to the Hawking formula. It shows that the temperatures of particularly massive Black Holes (e.g., for masses greater than a solar mass) remain below the ambient temperature of the Universe for all reasonable times, and thus the Black Holes initially grow. In the case that a Black Hole possesses the mass of the Sun, the situation reverses after about 10,000 times the present age of the Universe (i.e., after about 10^{22} seconds). The Black Hole would then still have 6.7×10^{74} seconds time to decay. For even more massive Black Holes, such a reversal would occur at an even much later time. Here it is to be noted that such times are beyond all observational limits.

For comparatively small Black Holes, Hawking's radiation formula foresees a cataclysmic end. The radiated power becomes increasingly explosive and the remaining lifetime decreases exponentially. A mini-Black Hole of about 1000 tons "flares up" in about one minute (see Table 11.2), and, by virtue of $E = mc^2$, converts about 2.5×10^{16} kWh into radiation energy. For comparison, this is about 150 times the annual world energy consumption. The strong space curvature, which surrounds a Black Hole and which is not considered in the Hawking formula, does the rest and will convert even larger masses into radiation energy in even shorter times.

Short intuitive outline of the quantum field theoretical processes in the case of Hawking radiation:

First: The **quantum field theory** is mathematically a quite demanding theory for the description of microscopic processes. It represents an extension of quantum mechanics and belongs to one of the great theoretical achievements of the last century. It has been validated without exception and with astonishing precision by an immense number of experimental results. It also finds applications in the technical field, e.g., in light emissions, in solar cells, in lasers, in LEDs, in superconductivity, in computer technology, in data transmission and data storage, etc.

In the quantum field theory the vacuum is not an "empty nothing", but a state, to which one can arbitrarily assign an energy $E = 0$. The quantum world allows fluctuations around this value, which arise from the fact that virtual particle-antiparticle pairs are permanently created in such a way that one of the partners assumes an energy greater than zero, the other one an energy smaller than zero, so that in the time average the energy remains at $E = 0$. Such pairs can consist of both massive and massless particles (e.g., photons), or of charged (e.g., electrons/positrons) or uncharged particles (e.g., neutrinos/anti-neutrinos). In the case of charged particles, the charges must add up on average to a zero charge. An outsider who lives in a macroscopic world does not feel anything about these processes happening in this complex world of the quantum vacuum.

If such a continuous process takes place close to the interface of a Black Hole, i.e., at the event horizon, the two partners can be separated, one being absorbed by the Black Hole, the other escaping into free space. In Fig. 11.16 this is shown.

If the particle that falls into the Black Hole carries negative energy, its partner (or anti-partner) has the corresponding positive energy, which can then be seen as a thermal radiation emanating from the surface of the Black Hole. The Black Hole, on the other hand, loses this energy (thus mass) and "shrinks". In the opposite case, the Black Hole absorbs energy, and the Universe cools accordingly. Which process finally dominates depends on the size of the Black Hole and the curvature of space. The relevant calculations show that mini- and micro-Black Holes explosively convert their far from negligible residual masses into radiation energy (see e.g., Table 11.2, columns 5 and 6.)

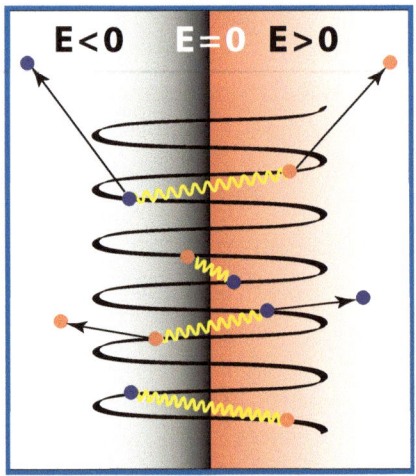

Fig. 11.16: Intuitive representation of vacuum fluctuations: At a boundary layer, a particle can separate from its anti-partner. If this separation occurs at the boundary to the Black Hole, one partner can be absorbed by the Black Hole with negative energy, while the other escapes into the "real" world with positive energy. This process occurs explosively for Black Holes at the end of their lifetime.

Primordial mini-Black Holes

Unfortunately, most of the theoretically postulated properties of Black Holes have not yet been verified experimentally. So-called primordial mini-Black Holes could bring light into this darkness — so go the wishes. According to theory, such objects would have formed shortly after the "Big Bang". Conditions during this early cosmic epoch would allow this, but only within a narrow time window of about 10^{-24} seconds after the "Big Bang" and also only <u>after</u> the inflationary phase[9]. The masses of such primordial

---

[9] Jacob D. Bekenstein, *BLACK HOLES: PHYSICS AND ASTROPHYSICS. Stellar-mass, supermassive and primordial black holes*, arXiv:astro-ph/0407560v1 (2004).

mini-Black Holes would then be distributed from 1 kg upward to about 1 billion tons (often compared to the size of a mountain). The lifetime of a 1 billion ton Black Hole, as can be seen from Table 11.2, is considerably longer than the present age of the Universe; one should therefore expect that the final and explosive "evaporation" of the less long-lived and thus smaller objects (initial mass: $\lesssim 170$ million tons, see Table 11.2) ought to be observable both, at present times and in all past epochs up to the times of cosmic background radiation — although one would have to consider that the smaller the original object, the more it lies in the past thus the fainter and more redshifted signals would have been emitted. There were many suspected sightings in the past years, but in the end none of these signals stood up to such an interpretation. Also scanning the cosmic background spectrum yielded no results. Nevertheless, the search for such mini-Black Holes continues unabated.

Undoubtedly, a positive signal would generate an enormous boost in knowledge. Not only that the emission spectrum of a Black Hole reveals a lot about its interior during its final stage, but also the mere existence of such primordial mini-Black Holes could be a hint that the super-massive Black Holes, which are found in the center of all giant galaxies, have evolved from such mini versions. These could then have triggered the development of the special structures of the galaxies already at a very early point in cosmic history. In fact, there is so far no conclusive explanation for the seemingly universal occurrence of such super-massive bodies like galaxies.

Rotating Black Holes

Rotating Black Holes have a particularly diverse and multifaceted physics. In 1963, the New Zealand mathematician Roy Kerr was the first to derive the Kerr geometry (or Kerr metric) for rotating Black Holes from Einstein's field equations [10]. It subsequently brought an enormous insight into the physics of Black Holes and into the observed phenomena in their vicinity. Some of the most remarkable findings will be touched upon in the following paragraph.

---

[10] Roy P. Kerr, *Gravitational Field of a Spinning Mass as an Example of Algebraically Special Metrics*, Physical Review Letters 11, 237 (1963).

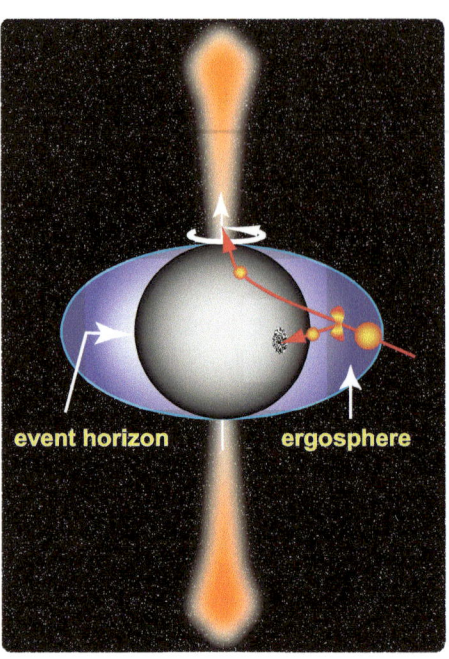

event horizon   ergosphere

Fig. 11.17: A rotating Black Hole is surrounded by an ergosphere. This is a region, where the co-rotating inertial frame created by the Black Hole cannot be compensated by a counter rotation, in such a way to make the surrounding starry sky appear stationary. An object, which penetrates into this area, can be torn apart as shown in the picture, whereby the one part with negative energy falls into the Black Hole, and the equivalent positive energy is transferred to the other part, which then escapes at the poles of the Black Hole with near light velocity (observable as the emission of a relativistic jet). This effect, known as the Penrose process, successively removes angular momentum and energy from the Black Hole. In the limiting case of extreme rotation, the radius of the event horizon (Kerr radius) is equal to half the Schwarzschild radius of the non-rotating system. In the region of the equator the other half is taken over by the ergosphere.

The rotation and the resulting centrifugal forces reduce the Schwarzschild radius, or synonymously the radius of the event horizon of the Black Hole. However, this is counteracted by the force-free inertial system dragged along by the Black Hole. The interplay of both effects leads to the formation of a so-called ergosphere, which surrounds the Black Hole in toroidal form (see Fig. 11.17) and which provides a window for observing the physics from the outside. An infall into the ergosphere is roughly equivalent to an infall into the Black Hole, but when falling into the ergosphere, there is a distinct possibility to escape again — though not without suffering significant damage.

If an object crosses the boundary to the ergosphere, it is from now on forced to co-rotate with the Black Hole, even if the entry was counter to the direction of rotation of the Black Hole. The twisted space-time forces the reversal of motion. The object perceives this situation via a counter-rotating starry sky within the ergosphere. A rotation against the direction of the Black Hole, so that the starry sky appears stationary, is no longer possible within the ergosphere. This also pertains to light, which is led mainly to the poles following the twisted space-time. When the infalling object reaches the event horizon, its rotational velocity is identical to that of the Black Hole, which in the extreme case is half the speed of light.

The ergosphere exhibits another remarkable property. If an extended object collides with another object inside the ergosphere, or when an object falls into the ergosphere from outside and is then torn apart, one part may receive a negative energy and falls into the Black Hole, the other part takes over the energy difference and may now leave the ergosphere with an energy even higher than the one originally present. It is primarily ejected in the polar regions as indicated in Fig. 11.17. The process is possible because of the extreme twisting of space-time. It is called the Penrose process after Roger Penrose[11]. It has no immediate analog in classical Newtonian physics. Penrose was furthermore able to show that the energy gained in such a process is taken from the rotational energy of the Black Hole, thereby causing its rotational frequency to decrease with time. Based on experimental data, this phenomenon could indeed be confirmed in all aspects.

The Penrose process is also one of the most relevant mechanisms for describing the physics of Active Galactic Nuclei (AGN). AGNs are the most luminous objects in the cosmos. They were discovered as early as the beginning of the last century, without scientists having even a rudimentary explanation for these phenomena on hand. In the 1960s, these enormous energy sources could finally be assigned to the centers of galaxies and later to the super-massive Black Holes they contain. The luminosity arises from the accretion of matter, which reaches relativistic velocities as it falls into the Black Hole, heats up through compression, and increasingly glows in the X-ray region before it all finally disappears into the ergosphere of the rotating Black Hole. By the Penrose process, which can further be fueled by the twisting of extreme magnetic fields (the Blandford-Znajek

---

[11] R. Penrose and R. M. Floyd, *Extraction of Rotational Energy from a Black Hole*, Nature Physical Science 229, 177 (1971).
Sir Roger Penrose, University of Oxford, UK, received the Nobel Prize in Physics in 2020 for his work on Black Holes. He shared the prize with Andrea Ghez and Reinhard Genzel.

mechanism [12], which is not further explained here), high-energy and narrow jets of energy and matter are ejected at the poles of the Black Hole. An instructive example is the jet from the super-massive Black Hole, which resides at the center of the M87 galaxy (see Fig. 11.18). The jet protrudes more than 5000 light-years from the center of the galaxy into the intergalactic gas causing it to glow.

The Black Hole in the Milky Way

The Milky Way, when judged by size and mass, is an average galaxy, and its Black Hole at the center with about 4.2 million solar masses in size, is even below average. Why its growth rate has remained so small in the cosmic past remains a question still to be answered. In Tab. 11.2 one sees that the Schwarzschild radius amounts to approximately 18 solar radii, which is just about at the edge of an optically resolvable dimension from a distance of (26.660 ± 85) light-years [13]. Unfortunately, the center of the Milky Way is obscured by interstellar dust and gas clouds and is therefore opaque to optical wavelengths. Yet, in the radio and infrared wavebands today's telescopes allow the central region with its Black Hole to be studied in detail. Here, the European Southern Observatory (ESO) in collaboration with several other scientific institutes have achieved a significant technological advance in instrumentation, which has made high-resolution observations of the Milky Way center possible. A surprisingly high density of stars was observed, and in the immediate vicinity of the Black Hole (here within a few light-years) the orbits of a large number of them could be determined with a previously unattained precision. The positions, which were recorded over 26 years with day-precision, revealed an extraordinary and almost chaotic motion in this inner region. Stars cavorting in all kinds of orbits around the Black Hole with at times extreme speeds (up to several percent of the speed

---

[12] R. D. Blandford, R. L. Znajek, *Electromagnetic extraction of energy from Kerr black holes*, Monthly Notice of the Royal Astronomical Society 179, 433 (1977)
The complicated equations of magnetohydrodynamics in the Kerr space-time can be solved today with powerful computers. The results are in general agreement with all observational data.
[13] The GRAVITY Collaboration, *A geometric distance measurement to the Galactic center black hole with 0.3% uncertainty*, Astronomy & Astrophysics 615, L10 (2019).

Fig. 11.18: The M87 galaxy in the constellation Virgo with its central super-massive Black Hole ejecting a jet of matter at relativistic velocities and making the interstellar gas glow. The jet in the image spans about 5000 light-years. The galaxy is about 53.5 million light-years away and brings about 10^{14} solar masses on the scale. The Black Hole in its interior has a mass of about 6.5 billion (6.5×10^9) solar masses. [Credit: NASA, public domain]

of light) were individually identified, and from their orbital parameters the existence of the Black Hole with a mass of (4.23 ± 0.11) million solar masses could be inferred without any doubt [14]. The Black Hole was also given the permanent name «***Sagittarius-A****» or «***Sgr-A****».

Figure. 11.19 shows a digital-image frame from an ESO video clip detailing the orbits of stars rotating around the Black Hole Sgr-A* at the center of the Milky Way.

In the course of this project, special attention was paid to the star S2 (see Fig. 11.19). S2 orbits Black Hole Sgr-A* within 16 years, 16 days and 19 hours on an extremely eccentric orbit and comes in its perihelion to within 19 billion kilometers from the Black Hole (about 17.5 light-hours). This is about 1500 times the Black Hole's Schwarzschild radius. At this point S2 reaches a velocity of 2.56% of the speed of light.

The orbit of S2 has been recorded with high precision for more than 25 years, and in May 2018 its perihelion could be observed directly for the second time. Special preparations had been made for this event just to be able to accurately determine the angle of perihelion rotation (see Fig. 11.19), as it is caused by the space curvature around Sgr-A*. One may recall, the space-curvature-induced perihelion rotation of the planet Mercury was 43 arcseconds ($= 0.012$ degrees) per century or 2.9×10^{-5} degrees per orbit of 88 days. In the case of S2, its perihelion rotation was determined to $\varphi = 0.22 \pm 0.04$ degrees per orbit; this is nearly 8000 times larger than the perihelion rotation of Mercury at a nearly 300 times greater distance from the central mass in the perihelion. The value calculated by General Relativity is $\varphi = 0.202$ degrees [15].

This result by the ESO collaboration is considered the first experimental

---

[14] During the writing of this book, the GRAVITY collaboration published new and more accurate data. The mass of the Black Hole was determined to be 4.297 ± 0.013 million solar masses and the distance to the center of the Milky Way was determined to be $26,989 \pm 30$ ly. (Cf. Gravity Collaboration, *Improved GRAVITY astrometric accuracy from modeling optical aberrations*, Astrononmy and Astrophysics 647, A59 (2021)).

[15] GRAVITY Collaboration, *Detection of the Schwarzschild precession in the orbit of the star S2 near the Galactic centre massive black hole*, Astronomy & Astrophysics 636, L5 (2020)
see also same authors, same title, arXiv:2004.07187v1 [astro-ph.GA], (2020).

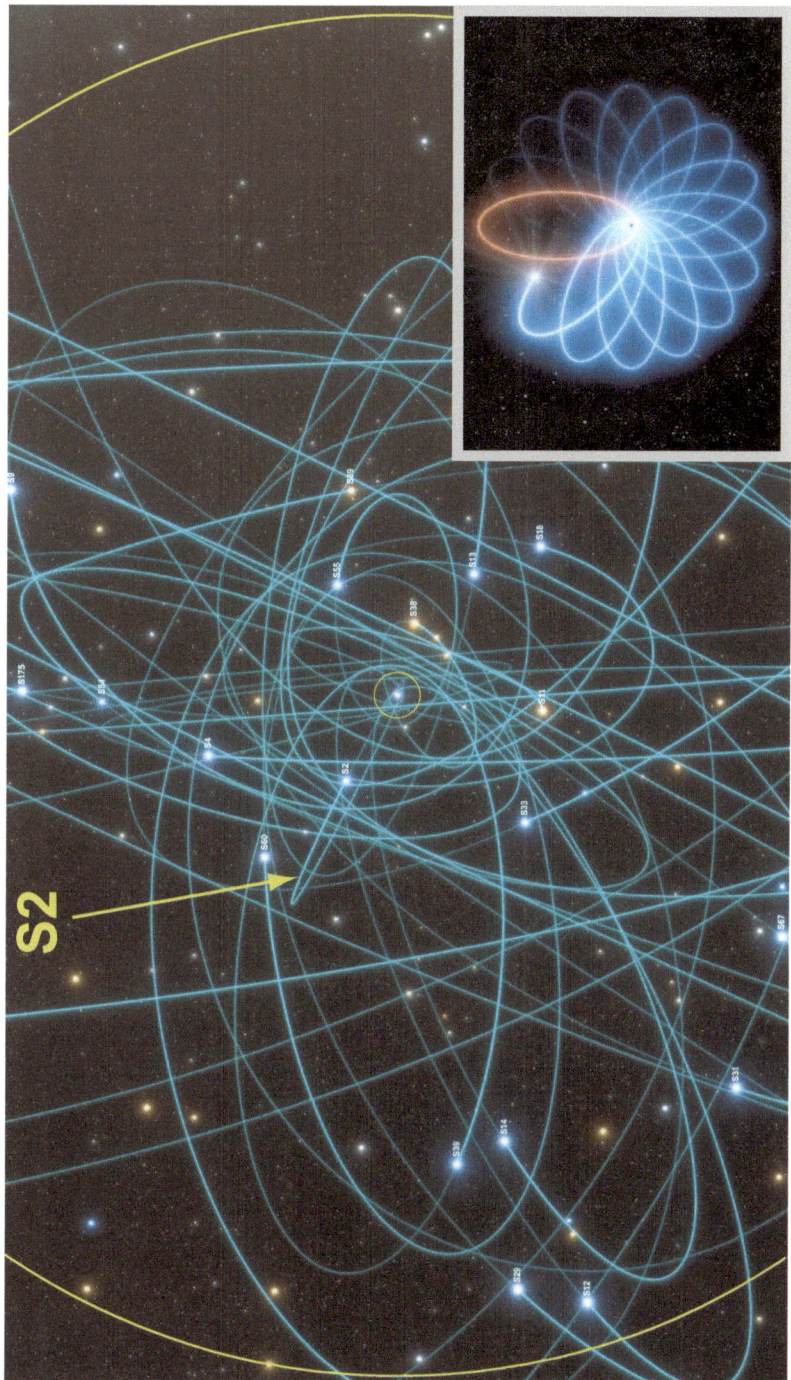

Fig. 11.19: A digital video-frame from an ESO video (id:1825f) detailing the star positions and orbits near the Black Hole in the Milky Way Center. The outer yellow circle represents the distance of a light-month from the Black Hole (in the center of the image), the inner yellow circle shows the distance of a light-day (about 26 billion kilometers). The star S2 is on an extremely eccentric orbit, whose perihelion rotation was studied with special attention to test the theory of General Relativity. — The lower right shows an artist illustration how the perihelion of a star rotates around the Black Hole, though exaggerated for better visualization. For a complete perihelion rotation S2 needs 1714 revolutions or about 27,500 years. [Credit: ESO/L. Calçada/spaceengine.org]

test of General Relativity in the case of extreme gravitational potentials. It allows an important conclusion: General Relativity not only correctly describes the tiny effects as they occur in our immediate environment, but is equally applicable in cases of extreme conditions. No changes are required so far.

These truly seminal achievements from this long-term project were awarded a Nobel Prize in Physics in 2020[16]. The interested reader will find further information on the subject of General Relativity and its verification on the ESO homepage.

The Black Hole in M87

The afore-mentioned M87 galaxy is an elliptical giant galaxy. It is located in the center of the Virgo cluster, which hosts a total of about 200 giant galaxies and 2000 galaxies of medium and small size. Among these, M87 is the largest and most massive, with an estimated total mass of about 10^{14} solar masses. Its distance from the Milky Way is about 53.5 million light-years. Because of its sheer size of about 5 million light-years in diameter, its unusual, nearly spherical symmetric shape, and especially because of its proximity to "us", it is a central object of research in astronomy and astrophysics.

M87 holds another record. In the center resides a Black Hole, whose mass of nearly 6.5 billion solar masses dwarfs everything else. How it gained this enormous mass, comparable to that of a medium-sized galaxy, is unclear. It is surmised that in the distant cosmic past two or more galaxies might have collided, from which M87 was formed. In this process also the Black Hole may have received a considerable part of the available mass.

In April 2019, the "**E**vent **H**orizon **T**elescope" (EHT) collaboration released the image of the Black Hole of the M87 galaxy shown in Fig 11.20[17]. It

---

[16] Andrea Ghez, Dept. of Physics and Astronomy Univ. of California Los Angeles, and Reinhard Genzel, Max-Planck-Institut für Extraterrestrische Physik München, Nobel Prize in physics 2020. They shared the prize with Sir Roger Penrose.

[17] The Event Horizon Telescope Collaboration, *First M87 Event Horizon Telescope Results. IV. Imaging the Central Supermassive Black Hole*, The Astrophysical Journal Letters 875:L4, (2019), https://doi.org/10.3847/2041-8213/ab0e85.

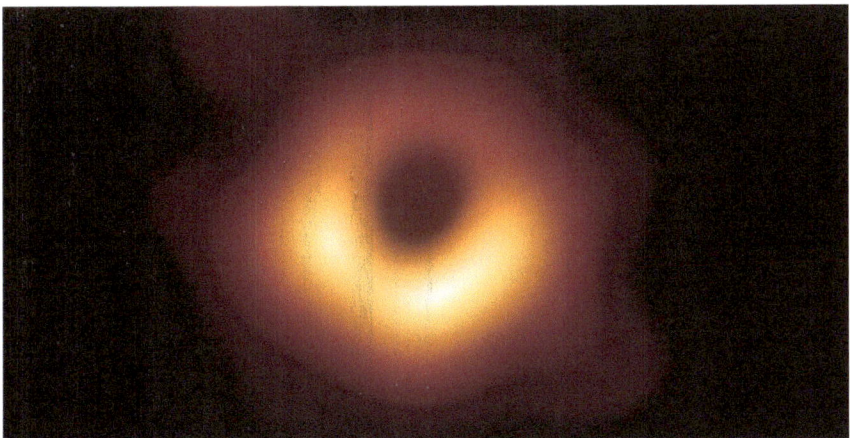

Fig. 11.20: Digital image of the Black Hole in the galaxy M87 created by the *"Event-Horizon-Telescope"* collaboration from measured data. As light coming from all directions is bent around the Black Hole, the central region appears darker to a distant observer. It is irrelevant from which direction the observation is made — the Black Hole is, so to speak, in its own shadow, independent of the observation angle.

shows a dark circle surrounded by a luminous aura created by the accretion of mass during the accelerated infall into the Black Hole. It is important to note that this image is not a photograph of the Black Hole. Rather, it is a visualization generated by computer from a large set of measurement data with a wide variety of identifiers. Several groups of the EHT collaboration had independently subjected the data material to a so-called "blind analysis"[18]. All groups eventually obtained an identical or highly similar picture[17].

For further explanation of Fig. 11.20: The extreme curvature of space in the immediate vicinity of a Black Hole causes light rays to bend around the

---

[18] In a "blind analysis," all incoming measurement data that could be used for later analysis are initially locked in a virtual data box — no one has access to these data. Meanwhile, computer programs are being developed and tested with computer-simulated data. The simulations take into account all the individual features of the detectors and their performance characteristics under different observational conditions. Neural networks — a commonly used software tool — are thoroughly tested and trimmed. The whole process can take several years. The data box containing the real measurement data is finally opened and made available to several software teams for their specific analysis. At this point, changes to the analysis software packages are no longer possible ("blind analysis"). Knowledge exchange among the teams is also prevented until all teams have completed their analysis.

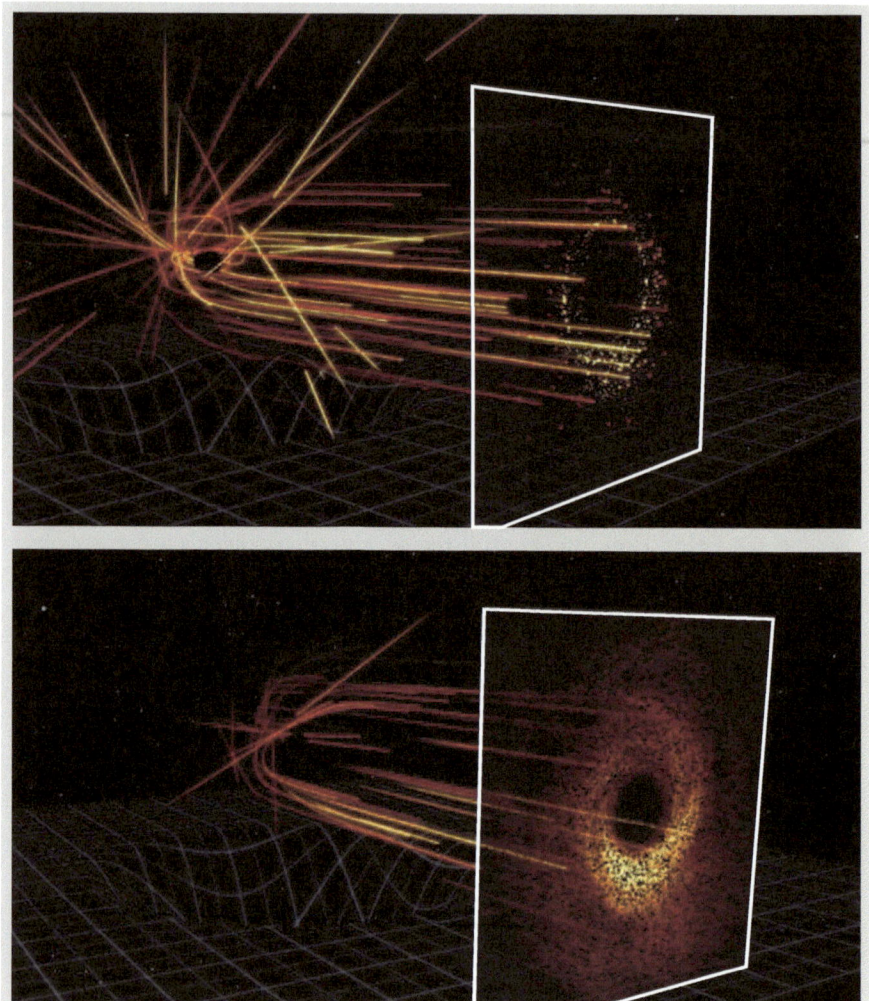

Fig. 11.21: Two frames of a video clip from the EHT collaboration to illustrate the image shown in Fig. 11.20. The increase in intensity at the bottom of the two images is attributed to the rotation of the Black Hole in the M87 galaxy.

Black Hole. By this intense lensing effect, light coming from any direction is focused onto a distant observer so that the periphery of the Black Hole appears bright and the center dark. Magically, it does not matter from which angle such an observation is made, the Black Hole appears always in its own shadow. This strange situation is visually illustrated in a video clip on the homepage of the EHT collaboration. Two frames are taken

from this video and shown in Fig. 11.21. They are intended to help explain
how the ring evolves around the Black Hole. And yet another observation
deserves attention at this point. The images show a significantly increased
luminosity on one side of the Black Hole. This is interpreted as an indication
of a rotating Black Hole. Infalling matter, and thus the accompanied light
emission, co-rotates with the Black Hole, which leads to an increase in
luminosity and to a Doppler shift of the light to shorter wavelengths on the
side moving towards the observer.

Between sense and nonsense

The fact that many aspects of the physics of Black Holes is not or only
incompletely known, and that many of the existing theoretical models
severely lack verification simply because of lack of experimentally secured
data, seems to be the cause for at times wild and abstruse speculations and
end-time fears, or for being highly dismissive of Science in general. This
attitude can be observed across entire societies. Here some things worth
knowing:

▶   Considering the age of the Earth or even the solar system and their
undisturbed evolution over cosmic times, a collision with a Black Hole,
which would annihilate Earth and/or the solar system just at a time when
the physics of these objects is being studied, is not particularly likely. Yet, it
is true that the formation of a Black Hole within a galaxy is not necessarily
a rare event on a cosmic time scale. The discovery of gravitational waves
in 2015 was due to a collision of two gravitationally bound Black Holes of
35 and 29 solar masses, respectively, at a distance of about 1.4 billion light-
years. In a mere 0.2 seconds three solar mass equivalents were converted
into gravitational wave energy, before finally a new Black Hole with only
about 60 solar masses remained. In the Milky Way, however, there is so
far no evidence that such a collision, or a similar one, ever happened in the
past, despite of ever improving observational techniques.

▶   Rotating Black Holes glow and emit high-energy jets. But even if a
Black Hole does not rotate, it does not necessarily escape direct observation,
because the luminosity of the continuously infalling matter is still visible
over cosmic distances.

▶   Science knows very little about primordial mini-Black Holes, except that they might exist. Here one will have to wait for an unambiguous observation, i.e., for a decay of such an object. The signature for this would be a sudden and explosive point-like source in interstellar or intergalactic space, with the spectrum of the emitted radiation corresponding to that of a blackbody at a characteristic temperature. All exploding mini-Black Holes should have the same characteristic temperature.

▶   A hitherto unmentioned property of Black Holes causes headaches to everyone who deals with them: If a Black Hole simply "evaporates" thermally, as envisaged by Hawking, an "information loss paradox" (or entropy problem) occurs. All matter and all pieces of information falling into a Black Hole will be annihilated, and everything eventually evaporates in the form of blackbody radiation, or equivalently, irretrievably dissolves into thermal random chaos. The entropy, as the quantity of disorder or randomness in thermodynamics, increases by up to 100 orders of magnitude, depending on the mass of the Black Hole, and this without any recognizable external influence. It is as if a well-ordered household falls into complete chaos from one minute to the other purely by chance and without any special intervention. Even the "Big Bang" releases information again and some kind of order re-establishes itself, otherwise we would not exist.

This fundamental problem of Black Holes was of course recognized also by Stephen Hawking in his theory, but without him finding a solution for it.

▶   In 2008, the Large Hadron Collider (LHC) was put into operation at CERN. Fears made the rounds that at the expected energies of $10^{12} - 10^{13}$ eV micro-Black Holes would be created with totally unforeseeable consequences. Apart from the fact that one has well-founded knowledge from nuclear physics about the enormous energies needed to compress an atomic nucleus even by less than 1%, let alone by a factor of 10^{40} (see Table 11.2), such a micro-Black Hole would instantaneously decay again. Also, the attractive gravitational potential emanating from it would be too small by about $30 - 40$ orders of magnitude to be even sensed by an object at a distance the size of an atomic nucleus. Moreover, the LHC energies are still at least a factor of 100 million smaller than the highest measured energies of particles that are part of cosmic rays and incessantly bombard the Earth. These measured energies are about $10^{20} - 10^{21}$ eV, and particles with such energies hit the Earth's atmosphere about 500 times per hour [19]. On the other hand, if we take 10^{13} eV, i.e., LHC energies [20], as the lower benchmark energy, we find that the cosmic particle flux onto the Earth's atmosphere is about 3000 billion particles per <u>second</u>.

Micro-Black Holes that have swallowed the Earth have not occurred in the last 4 billion years.

---

[19] "The Pierre-Auger Collaboration", *Features of the Energy Spectrum of Cosmic Rays above* 2.5×10^{18} *eV Using the Pierre Auger Observatory*, Physical Review Letters 125, 121106 (2020).

[20] The LHC energies quoted here are energies/nucleon. Only these values are relevant here.

Chapter 12

Cosmic Messengers

The discovery of cosmic rays
and the particle components that
constitute them, together with the
extensive research, which emerged from it,
is certainly one of the most astonishing and also most ground-breaking
chapters of physics. The spirit of research associated with it, as well as the
unflagging efforts of the persons acting in this field is almost unsurpassed
until today. Since its beginning about 100 years ago, there was probably no
other field of physics, which experienced so many surprises, and which so
often left researchers in disbelief — and this up to the present time.

It is particularly remarkable that this research field began with the simplest
instruments and tools. These were initially the size of a shoe-box and, yet,
were nevertheless able to deliver fundamental and entirely new insights. For
comparison: The size of today's experiments for the investigation of cosmic
rays, such as the Pierre-Auger Observatory in Argentina, have an areal
dimension of nearly $3000\,\text{km}^2$, or the IceCube detector at the South Pole, a
volume size of about $1\,\text{km}^3$ — that compares to about 2×10^{11} shoe-boxes.
In the first case, the approximately $50\,\text{km}$ thick atmosphere is the medium
into which the cosmic rays strike, and in the latter, it is the approximately
$3000\,\text{m}$ thick ice layer of the Antarctic ice cap, where cosmic neutrinos,
which have already been filtered out by the atmosphere, leave their signal.

© The Author(s), under exclusive license to Springer-Verlag GmbH, DE, part of Springer Nature 2025
D. Frekers, P. Biermann, *Universe, Neutrinos, Stars and Life*, https://doi.org/10.1007/978-3-662-70729-6_12

12.1   History

The discovery of cosmic rays was by no means a goal-oriented research project. At that time not even the term «*cosmic radiation*» existed; this was coined only as late as 1925 by the American physicist Robert Millikan (1868 – 1953).

Preceding the discovery was an unexplained phenomenon that had been known to exist since the mid-18th century.

To remember: *The mid-*18th *century was the time, when research on electricity started. However, the terms electrons, protons, atoms, photons and also the chemical elements (classified only in 1869 by Mendeleev and Meyer) did not yet exist.*

It was Charles-Augustin de Coulomb (1736 – 1806) who observed in his capacitance experiments on electrostatics that electricity, or electric charge, always disappeared in an inexplicable way[1]. Even the best insulators could not stop this disappearance. By the end of the 19th century, the prevailing notion was that electricity "disperses" according to a generally accepted law, i.e., the "Coulomb law of dispersion". All experiments up to that time were consistent with this law.

It was not until 1887 that W. Linss[2] found in his meteorological and air-electrical studies that "Coulomb's law of dispersion" contained a very significant weakness. In Coulomb's conception, charges should be carried away primarily by aerosols in moist air. Linss, however, found the exact opposite in his remarkable experiments, which he conducted continuously over two years: It was dry air, not moist air, which possessed the higher electrical conductivity, as dry air "dispersed" electrical charges more rapidly. This realization set a multitude of similar meteorological and air-electrical investigations in motion, which led, among other things, also to the conclusion that an electric field emanated from the Earth's surface. But where were the opposite charges, towards which the field lines were heading, and where were the obviously never-ending charge reservoirs, which sustained

---

[1] C. A. Coulomb, Mémoires de l'Académie de Paris, S. 616 (1785).

[2] Dr. Linss, *Ueber einige die Wolken- und Luftelektricität betreffende Probleme*, Meteoro- logische Zeitschrift 4, 345 (1887) and Dr. W. Linss, Darmstadt, *Ueber Elektricitätszerstreuung in der freien Atmosphäre*, Elektrotechnische Zeitschrift 38, 506 (1890).

the electric current that seemed to be constantly emanating into the conducting air?

A note: *Today, it is known that the difference of the electric potential between Earth and atmosphere amounts to about 100 to 300 V per meter height difference near the surface, but drops significantly at high altitudes. If one takes an average conductivity of the atmosphere, the migration of the free charge carriers results in a constant vertical current, which amounts to about 500 A summed up over the entire surface of the Earth, thereby leading to a surplus of positive charges near the surface. Lightning strikes eventually provide the discharge to keep the electric potential constant.*

At those times, though, the most divergent opinions circulated, which were discussed highly controversially and even fiercely at times, e.g., that the clouds were electrically charged, or the field lines extended up to the Sun or even up to the infinity of the cosmos.

A note: *That electrons were carriers of electric charge was not recognized until 10 years later, around 1897, by Sir J. J. Thomson[3]. It was he, who coined the term «electron».*

Remarkably, Linss already suggested in his work to let kites rise to higher altitudes and then determine the potential gradient of the electric field at different points in the atmosphere by use of a specially constructed "on board"- electrometer and to do this under different weather conditions. It was not until more than 10 years later that Victor Hess took up this idea again and realized it, for which he was subsequently awarded a Nobel Prize.

However, a lot had changed during these 10 years:

▶ More or less accidentally, Henri Becquerel discovered radioactivity[4] in 1896. He was experimenting with the phosphorescence of uranium salts and found that a photographic plate, on which he had deposited some samples, had blackened despite being wrapped. This discovery, though, did not gain significant attention.

---

[3] Sir Joseph John Thomson (1856–1940), Nobel Prize 1906 for his research on the electrical conductivity of gases.

[4] Henri Becquerel (1852–1908), Marie Curie (1867–1934) and Pierre Curie (1859–1906), Nobel Prize 1903 for the discovery of radioactivity. Marie Curie, née Sklodowska was a student of Becquerel and the first woman to be honored with a Nobel Prize in Physics.

▶   A year later, in the course of his studies on cathode rays, Sir Joseph Thomson identified the electron as the carrier of electric charge, and

▶   around 1900, Becquerel again showed that the radiation escaping from the atomic nucleus was magnetically deflectable. However, that these so-called β-rays were merely fast electrons was not shown until 1948 by Maurice Goldhaber and Gertrude Scharff-Goldhaber [5],

▶   and finally Marie and Pierre Curie were able to use an electroscope to estimate the relative strength of the new radiation via the ionization of the air. They also showed that the conductivity (or "charge dispersion") of the air was related to the presence of charged ions, which arose as a consequence of the charge-separating effect (ionization) of the radiation. The term "radioactivity" was coined here.

With the discovery of other natural radioisotopes such as those of thorium, radium, or polonium [6], an even more pertinent question about the electrical conductivity of air was brought forward: up to what altitudes is the atmosphere still noticeably ionized by radiation, how intense is the ground activity, and are there perhaps other undiscovered natural emitters at the ground surface? However, the fundamental problem of the conductivity of air at altitudes outside the range of influence of the radiation remained, because even with effective shielding of the radioactive sources, a "background charge dispersion" of the air still remained. To investigate this problem, it was necessary to carry out measurements at higher altitudes above ground level, which at that time was considered an unrealistic undertaking.

A pioneering technical development in this context was the "two-filament electrometer" developed by Theodor Wulf [7] in 1906. For the first time, such a device had a convenient size (about that of a shoe box), and despite its small size it covered an extremely wide measuring range given by the standards of the time; it was sensitive to the smallest amounts of a voltage

---

[5] M. Goldhaber, Gertrude Scharff-Goldhaber, *Identification of Beta-Rays with Atomic Electrons*, Physical Review 73, 1472 (1948).

[6] The element polonium was named in honor of Marie Curie after her native country Poland.

[7] Theodor Wulf (1868–1946), born in Hamm in the province of Wesphalia, physicist and Jesuit priest.

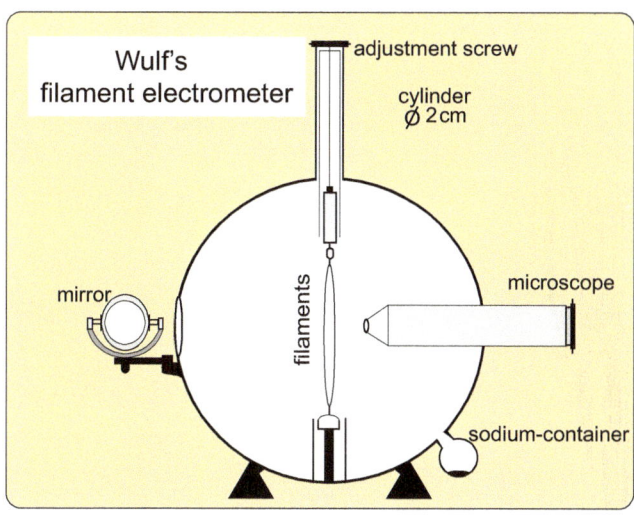

Fig. 12.1: Wulf's "two-filament electrometer". Two quartz filaments (thickness about $1-2\,\mu$m) slightly stretched with a clamp inside a cylinder, repel each other by applying an electric voltage. Through an opening cut in the cylinder jacket, the filaments are illuminated by a mirror and observed through a microscope on the opposite side. A discharge process reduces the distance between the filaments. A sodium reservoir removes moisture from the air and ensures reproducible conditions.

generating charge (a few scales per volt) and was also extraordinarily robust. It was charged by a dry cell battery. The literature of the time praised this electrometer as being "of a unique quality", and indeed this instrument was subsequently used for almost all measurements of air conductivity at a wide variety of locations on land and at sea with the results always being the same and consistent. Wulf himself (Fig. 12.2) made a trip to Paris in 1910 to make his own electrometer measurements at the height of the Eiffel Tower, which was at that time the tallest structure in the world. He was convinced that at 300 m the ionization effect due to ground radioactivity would have largely disappeared. Surprisingly though, he found a decrease of only about 10%, much less than the expected value of about 75%, seemingly indicating an unknown radioactive source within the atmosphere.

Wulf's "two-filament electrometer" is shown in Fig. 12.1 with a description of its mode of operation.

Fig. 12.2: Theodor Wulf (around 1910), physicist and Jesuit priest, developed the "Wulf electrometer", which led to the discovery of cosmic radiation.

Fig. 12.3: Victor Franz Hess, discoverer of the cosmic radiation.

Fig. 12.4: Victor Franz Hess (in the gondola in front) just prior to his balloon flight.
[Credit: © Victor-Franz-Hess-Gesellschaft e.V. – archive, permanent loan VFH-Breisky, Pöllau]

Victor F. Hess — an amazing career

It must have been pure adventurousness coupled with an unswerving pioneering spirit and a good dose of naivety that made Victor Hess, then just 27 years old and a fresh honorary professor at the University of Veterinary Medicine in Vienna, decide to finally get to the bottom of the problem of the atmospheric ionization using an entirely new approach. From his more medically oriented training, the physics of radioactivity and radiation was not exactly within his expertise. His proposal to undertake balloon flights in order to then carry out measurements of air ionization at higher altitudes could therefore hardly be classified as a leisurely afternoon excursion. Moreover, Hess was by no means an expert in the control of an open-air balloon. Amazingly, between 1911 and 1912, he managed to get nine balloon flights organized and financed, with the first two in 1911 being more or less training rides. Of the seven balloon ascents carried out in 1912, the last one was the most daring but also by far the most decisive. Four of these particularly spectacular flights shall be briefly commented upon here: these are the first, the fourth, the fifth and the last.

▶   The first flight in an open-air balloon took place on April 17, 1912 at about 10:30h in the morning from the Vienna City exhibition center, the Prater. The date is remarkable in so far as on this day a total solar eclipse took place over Western and Northern Europe, and also in Vienna a large part of the solar disk was still covered. Equipped with several electrometers on board, the goal was to determine the change in ionization of the air during the different phases of the eclipse, to see if the ionization was caused by the Sun. The balloon reached altitudes between 2000 and 2800 meters and was thus above the cloud cover that was closed on that day. The planned measuring program could not be carried out completely because during the eclipse the balloon cooled down and Hess and his balloon pilot were forced to an early landing about 135 km away from Vienna. Despite all this, a noticeable change of the air ionization during the eclipse of the Sun was not observed.

▶   The fourth flight was a night-flight during the night of June 3 to 4, 1912. After a series of measurements close to the ground, i.e., between 800 and 1200 meters altitude, Hess wanted to use the warming influence of the

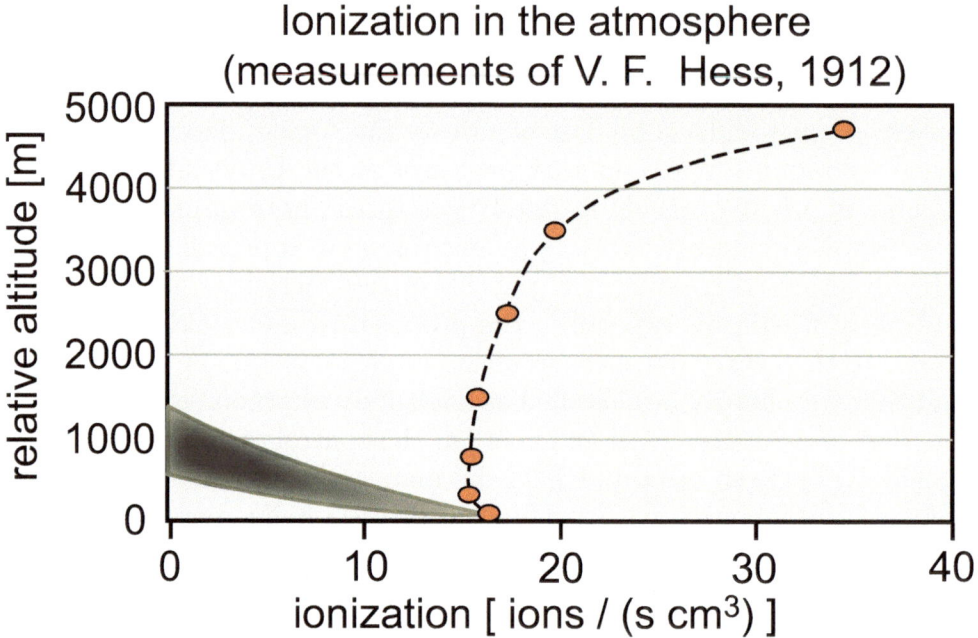

Fig. 12.5: The measurements (in red) of Victor Hess show that the atmosphere, quoting V. Hess: "from above", is exposed to intense ionizing radiation. The shaded area would approximately be the expected altitude-dependent ionization, which would have resulted from the natural radioactivity of the Earth's surface.

Adapted from: G. Federmann, *Viktor Hess und die Entdeckung der Kosmischen Strahlung*, dipl. thesis, Inst. f. Radiumforschung und Kernphysik, Univ. Wien (Jan. 2003). It also contains the values from the original measurement protocols.

early morning sun to advance to higher altitudes. For this purpose, Hess had borrowed a particularly large balloon with a capacity of 2200 m³. The plan failed; an early morning thunderstorm front forced the two-man crew to land.

▶ The fifth flight took place a few days later and was a solo-flight. Apparently, Hess's staff was getting too suspicious of his undiminished drive, so that Hess — this time left on his own — had to take over both, the ionization measurements and the control of the balloon. Hess reached a maximum altitude of 1200 meters.

▶ The seventh flight on August 7, 1912 was finally the decisive flight. Since Hess had so far not discovered any reduction in air ionization, but rather noticed a slight increase with increasing altitude, the ascent to much higher altitudes seemed imperative to him. His financial backers, however, viewed his activities with increasing suspicion and blocked the funds for further projects. Hess finally managed to get a balloon made available to him in Bohemia (at that time Austrian Crown-Land) but only for just one flight. The ballon was filled with $1680\,m^3$ hydrogen (!!). Equipped with additional meteorological instruments and air pressure gauges and three differently shielded electrometers, he and his crew of two finally ascended in Aussig on the Elbe River on the early morning of August 7. The wind carried the balloon northwest towards Berlin.

They advanced to an altitude of 5350 meters and made continuous measurements. Unfortunately, with increasing altitude, Hess developed the symptoms of an increasingly life-threatening altitude sickness, so that the flight had to be aborted after only 6 hours at noon. Yet, the measured data taken up to that point already painted a clear picture. From about 2500 meters altitude the data showed an unmistakable increase of the ionizing radiation and from about 4500 meters the radiation already exceeded that on Earth by almost a factor of two. This increase was registered by all three measuring instruments in a fully consistent manner. However, the last value above 5000 meters was accidently erased before read-out by an unfortunate movement of the meanwhile highly distressed Hess.

The data processed from the original protocols are shown in Fig. 12.5. They confirm the statement of V. Hess, *"... **that radiation of very high penetrating power enters the Earth's atmosphere from above"***

Victor Franz Hess (Figs. 12.3, 12.4 and 12.6) received the Nobel Prize in physics for this work in 1936, which he shared with Carl David Anderson (see following pages).

Fig. 12.6: Victor Franz Hess (1915); next to him, Wulf's electrometer.
[Credit: Globetemp, CC BY-SA 4.0]

Marietta Blau — pioneer of modern astroparticle physics.

Marietta Blau is not particularly well known in the physics community — unfortunately! — because her discovery should be classified as even more significant than that of Victor Hess. Marietta Blau came to the Institute for Radium Research of the Austrian Academy of Sciences in 1923, i.e., about 10 years after Victor Hess made his discovery there. She worked as an unpaid scientist with photographic emulsions and investigated their suitability for the detection and identification of ionizing particles (mainly protons and α-particles) and particle tracks. In this effort, she worked closely and quite successfully with the Ilford Company and managed to significantly improve the manufacturing techniques as well as the homogeneity and fine-grainedness of the emulsions.

After the discovery of cosmic rays by Victor Hess, the branch of cosmic-ray science became a central and rapidly growing research topic, even worldwide.

In the US, Carl Anderson (1905–1991) discovered in 1932 the positron as the antiparticle to the electron in cosmic-ray showers, which he made visible in cloud chambers. Paul Dirac had predicted its existence 3 years before in his relativistic quantum theory, which is still unrestrictedly valid today.

Carl Anderson «*for the discovery of the positron*» and Victor Hess «*for the discovery of cosmic rays*» were awarded the Nobel Prize in Physics in 1936.

Marietta Blau (Fig. 12.7) suspected that the photographic plates, which she was able to make, could also be employed to study cosmic rays. She therefore placed several stacks of those plates for a period of five months on a mountain station, which V. Hess had already used, and which was located at an altitude of 2300 meters on the "Hafelekar" mountain near the city of Innsbruck. The subsequent microscopical analyses of the embedded tracks showed a truly unexpected result. The cosmic rays possessed components of highest-energy particles, whose energies were all enough sufficient to fully disrupt ("zertrümmern") the bromine and silver atomic nuclei in the emulsions after a collision.

Fig. 12.7: Marietta Blau, Austrian physicist (photo circa 1927). [Image source: Eva Connors; Image archive, "Zentralbibliothek für Physik" in Vienna]

Nuclear reactions in this extreme form were completely unknown at that time. The photograph published in 1937 by M. Blau of such a nuclear reaction, for which she coined the word "Zertrümmerungsstern – shattering star", is shown in Fig. 12.8. This remarkable picture set a completely new accent to the study of cosmic rays.

Photographic emulsions continued to be used in cosmic-ray research, especially because their quality gradually improved. The spatial resolution in the sub-micrometer range played a decisive role, and even today's detectors of high-energy physics use photographic emulsions at times. The key features when combined, give information about mass, charge, momentum and energy

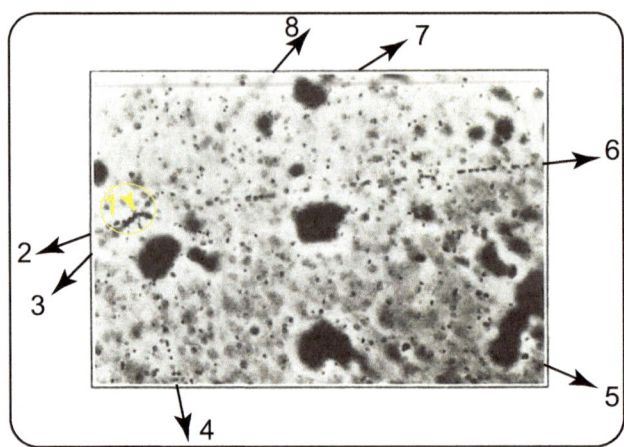

Fig. 12.8: First demonstration of the "shattering" of a bromine or silver nucleus in a photographic emulsion after a collision with a cosmic-ray particle. A total of 8 tracks are identifiable in this 70 μm thick photographic emulsion. Such a high multiplicity is not possible in a natural radioactive decay process. Because of the thickness of the emulsion plate, the tracks are not in the microscope's focus along their entire length. Track 1 impinges nearly perpendicularly and is only identifiable in a 3-dimensional scan.
This image fundamentally changed the previously existing perception of the cosmos. [Source: M. Blau, H. Wambacher, Nature, 585 (1937)]

of a particle traversing the emulsion. These are: (1) the track width, (2) the so-called δ-electrons, which are knocked out of an atomic orbit of the medium atoms and emitted perpendicular to the track, (3) the small angle changes caused by small-angle multiple scattering in the medium, and (4) the precisely determinable spatial coordinates of a particle track. Even short-lived, neutral particles can be identified by their decays or by their secondary reactions. With some analytical effort and with powerful microscopes an exact reconstruction of the primary reaction is therefore almost always possible.

Two examples of cosmic impacts of heavy atomic nuclei with energies of about 10^{13} eV are presented in Fig. 12.9. Both photographs were taken in the 1950s.

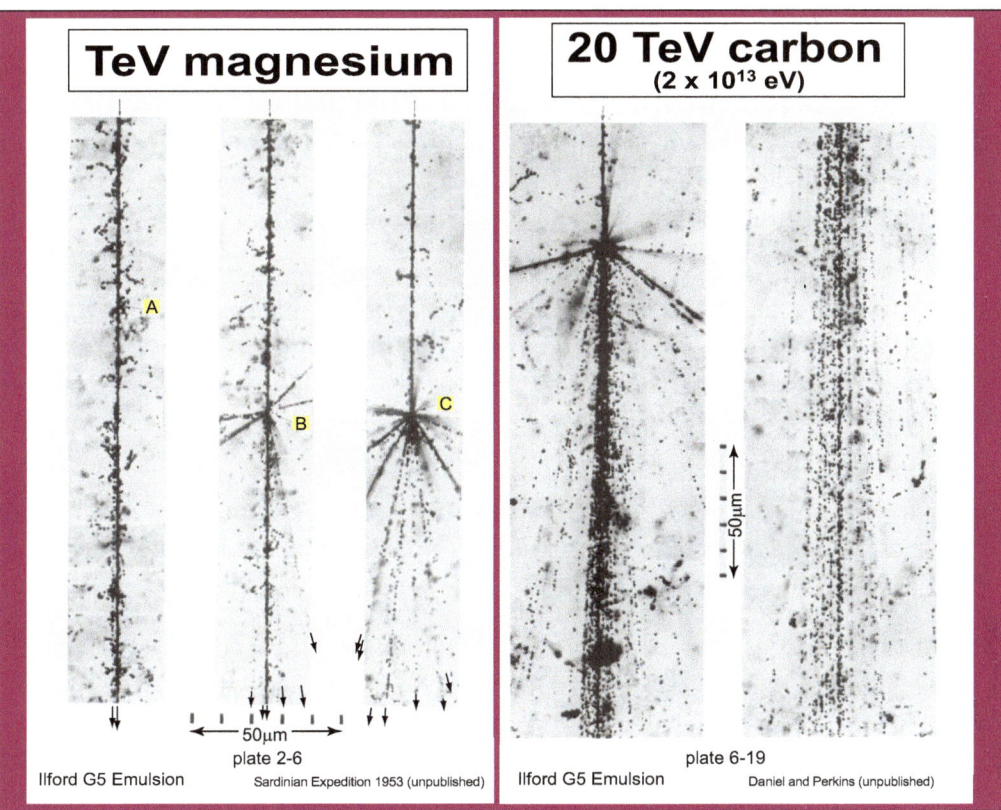

Fig. 12.9: Left: A high-energy cosmic-ray particle (identified as a magnesium or aluminum nucleus) penetrates a stack of photographic emulsion plates, leaving a characteristic track followed by 3 further nuclear reactions involving the atomic nuclei of the emulsion (silver or bromine) (points A, B, and C). At point A (3 cm after entering the stack), a peripheral collision is identified, where the cosmic particle "evaporates" an α-particle, thereby losing 2 charge units ($2e$). The two fragments move nearly parallel (see arrows at bottom). After about 2 cm (now in plate 4), another reaction occurs, producing protons and charged pions. The primary nucleus loses about 5 charge units here ($5e$). At point C (after about 1 cm again and in plate 6), a reaction occurs that leads to fragmentation of both the primary nucleus and the nucleus of the emulsion that was struck. Charged nucleons and pions are emitted in a star-like shape. Their tracks can be traced through 30 emulsion plates that follow. — **Right:** A carbon nucleus of cosmic origin at about 20 TeV collides with an atomic nucleus in the emulsion (bromine or silver). The collision leads to a complete disintegration of the two colliding objects, thereby producing more than 100 pions.

In an analysis the tracks are followed in the emulsions via labor-intensive microscopy. Particle and particle charge identification is performed by measuring the track widths and by determining the number of electrons emitted nearly perpendicular to the tracks (so-called δ electrons). The latter are emitted by the bromine and silver atoms, when the primary electric charge passes through, leaving short and non-straight tracks.

[Image source: *The Study of Elementary Particles by the Photographic Method, An account of The Principal Techniques and Discoveries illustrated by An Atlas of Photomicrographs* by C. F. Powell, P. H. Fowler and D. H. Perkins, Pergamon Press 1959]

Historical Addendum

A short addendum to these historical facts is appropriate at this point, especially since a theater team had staged a performance especially for the Münster Astro-Seminar, which took place in 2012 and which brought the history on this subject during the Nazi era onto a theater stage.

Note: *When individual fates are described at this point, the authors of this book in no way wish to exclude the suffering that fell upon so many people during the Nazi era. However, the authors believe that individual fates create a much more lasting effect and are therefore suitable for recalling the past more vividly and intensively.*

On March 13, 1938, the "Anschluß" – "annexation" of Austria took place, and with it the "Arisierung" – "Aryanization" of the country by the German Nazi dictatorship. This date represented a turning point in the lives of both Victor Hess and Marietta Blau.

Victor Hess had already openly and vehemently declared his opposition to the Nazi-regime. This did not remain without consequences. In September 1938, his position as a physics professor in Innsbruck, which he had held since 1931, and eventually as director at the new Institute for Radiation Research, was terminated, effective immediately, and without any subsequent pension entitlements. The confiscation of his Nobel Prize money was also ordered. Shortly before his pending arrest, he fled with his wife Maria Bertha, who was of Jewish descent, to Switzerland and from there to the United States, where he continued his scientific work at Fordham University in New York City, and where they both received US citizenship in 1944. His membership in the Academy of Sciences in Vienna was terminated in 1940, but he was re-appointed to the Academy as a Corresponding Member Abroad shortly after the end of the war. Victor Hess returned to Austria and Innsbruck only once. He died in 1964 in Mount Vernon, NY, USA.

For Marietta Blau[8], who also had Jewish roots, this meant a degradation of her person even before this date. It affected her throughout her life. At

---

[8] Information about Marietta Blau taken in part from: Robert Rosner, Brigitte Strohmeier (eds.): Marietta Blau - Sterne der Zertrümmerung. Biography of a Pioneer of Modern Particle Physics, Böhlau publisher, Vienna 2003.

the beginning of 1938 she accepted an invitation for a research stay at the University of Oslo, and when she left Vienna for Oslo on March 12 (a day before the annexation), she realized on the way that a return was no longer possible for her. In Oslo she worked in the chemical laboratory for a short time.

Through a personal letter of recommendation from Albert Einstein, in which he pointed out the "unusual talent" of Marietta Blau, she was offered a position at the Technical University in Mexico City. There she was to set up a research laboratory on cosmic rays. She arrived in Mexico by a circuitous route in October 1938 and began her work as a professor in January 1939, initially with great enthusiasm. The conditions in Mexico City were predictably poor and the monthly salary of $ 100 rather meager, and this amount was not always paid to her either. In addition, she complained bitterly about competitiveness and controversy and, above all, about the particularly pronounced male dominance that she had to endure as the only woman in the 50-member teaching staff.

In 1944, she finally immigrated to the United States and worked in industry until 1948. She then returned to academia, where she stayed until 1960. During this time she was engaged in the development and production of photographic methods for particle detection in high-energy experiments at various accelerators in the US, with great success.

She felt a particular bitterness that the work she had left behind in Vienna was used and also published by her colleagues at that time without even mentioning her, where, on the other hand, their loyalty to the Nazi regime was rewarded with high positions. She was also pained by the fact that Cecil Powell was awarded the Nobel Prize for Physics in 1950 for work that had been inspired to a large extent by her and her colleague at the time, Hertha Wambacher, but that Powell did not even mention this work in his Nobel Prize speech, although he was well aware of it. There was even a proposal, put forward by Erwin Schrödinger (Nobel Prize for Physics 1933) in early 1950, to share the prize between Powell, Blau and Wambacher. However, Hertha Wambacher died in April 1950. Hans Thirring (see footnote on page 225) proposed Marietta Blau again for the Nobel Prize in 1955, and Erwin Schrödinger also submitted a corresponding proposal again in 1960.

In 1960, Marietta Blau returned to Austria, where she again conducted research as a private employee with no pay at the Radium-Institute in Vienna until 1964. She analyzed photographic images of particle tracks from experiments at CERN.

In 1970, Marietta Blau passed away in Vienna, totally impoverished. An obituary for her in a scientific journal did not appear.

Her name was obliberated.

It was not until 2004 that a commemorative plaque was unveiled at the Gymnasium, Rahlgasse 4 (Fig. 12.10), where she took her graduation exams in 1914, and the City of Vienna finally gave the Marietta-Blau-Gasse (Marietta-Blau alleyway) in the 22nd district her name.

Fig. 12.10: Memorial plaque for Marietta Blau

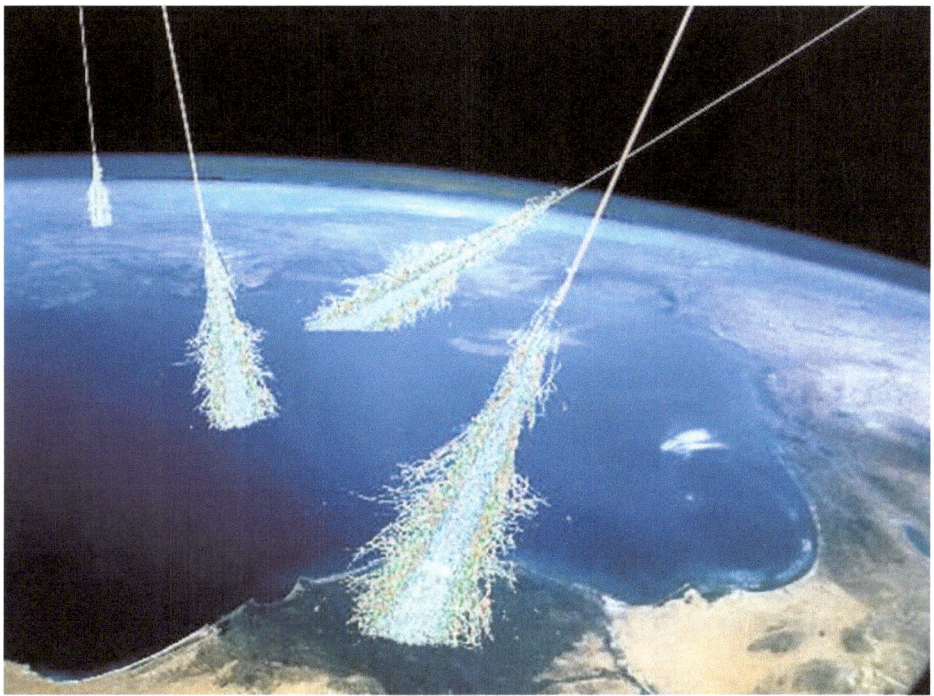

Fig. 12.11: Air showers generated by highest energy cosmic rays, — an illustration for clarification. [Credit: © Simon Swordy (U. Chicago), NASA]

12.2   Decoding the cosmic radiation

Cosmic radiation, which has been researched for more than 100 years now, has impressively shown how extremely hostile to Life the seemingly empty space outside the Earth is, but also how effectively the Earth protects its own habitat from this radiation. The Earth's magnetic field together with the atmosphere reaching several hundred kilometers high shields about 99.9% of the radiation. Especially the highest energetic particles "shower up" as soon as they hit the atmosphere, and they thereby distribute their energy by generating secondary particles over an impact area of up to several 100 km^2. Also nitrogen in the Earth's atmosphere performs an important shielding function, which will be discussed separately later. An illustration of the shower effect is shown in Fig. 12.11.

It was Pierre Auger[9] who tracked down this shower phenomenon as early as 1939. He showed that detectors, which he had placed up to 300 m apart, sometimes registered coincident events. He correctly interpreted this as a cascade of secondary particles generated by the primary cosmic ray in the atmosphere. The current largest-area air shower detector in the Argentinean Pampas therefore bears his name.

Since that time, cosmic-ray research has expanded at an almost unprecedented rate. Detectors have been placed underground, in glacial ice, in the deep sea, on the ground, or more recently aboard satellites, either in Earth orbit or on interplanetary missions. Even in lunar rocks and meteorite dust, traces of cosmic rays are embedded and identifiable. These objects are important witnesses because they provide information about changes in the intensity and composition of radiation as they may have occurred over cosmic time scales. It is beyond the scope of this book to describe all of these projects and their scientific merits individually. Therefore, after a brief description of the current experimental findings and a short digression on gravitational waves, we will limit ourselves to the three largest and perhaps most spectacular projects, namely the IceCube project at the South Pole, the aforementioned Pierre-Auger project in the Argentinean Pampas, and the TA project in the US state of Utah.

The questions of initial interest are:

▶ What is the spectrum of the cosmic radiation, and is there perhaps a maximum energy that cannot/must not be exceeded in the cosmos?

▶ What are the sources of the cosmic rays and where are they located?

▶ What physical processes produce the energies that occur especially in the extreme high-energy range of the cosmic radiation?

▶ What particle components make up the cosmic radiation?

▶ Is the cosmic radiation subject to short-time changes that can be correlated with other observable cosmic events?

---

[9] Pierre Auger, et al., *Extensive Cosmic-Ray Showers*, Reviews of Modern Physics 11, 288 (1939).

▶ How can the observed isotropy of the cosmic radiation be understood?

The experimentally determined energy spectrum of cosmic radiation is shown in Fig. 12.12. It covers a range of more than 35 orders of magnitude, which is something rarely seen in experimental physics. It summarizes data from a large number of independent experiments conducted at a wide variety of points on Earth. In addition to its scientific relevance, this spectrum is also evidence of a well-functioning, world-wide cooperation in this field.

The double-logarithmic representation only insufficiently brings out the subtleties in the spectrum. The energy dependence follows a relatively steep slope with an $E^{-\gamma}$ power law with an initial value of the exponent $\gamma = 2.6$, which changes to even larger values from about 10^{15} eV onwards, corresponding to an even steeper decreasing power law of the energy dependence. In this double-logarithmic representation (Fig. 12.12), the structures designated in a technical jargon as "knee" and "ankle" appearing at about 10^{16} eV and $10^{18.5}$ eV are only barely visible.

> It may be useful at this point to draw a parallel to the energies occurring here. A particle energy of 10^{20} eV (10^{11} GeV) is equivalent to the energy of a tennis ball of about 90 km/h.

In the Figs. 12.13 (a) and (b), the spectrum from Fig. 12.12 is therefore multiplied by $E^{+2.6}$, which makes the embedded structures appear more clearly. Further plotted in Fig. 12.13 (a) are the values from two key projects, i.e., PAMELA and CREAM-II, which give information about the particle composition of the cosmic radiation.

The PAMELA (**P**ayload for **A**ntimatter **M**atter **E**xploration and **L**ight-nuclei **A**strophysics) detector was launched into Earth orbit in 2006 aboard a Russian Earth-observation satellite. The primary goal was to use this device to detect antimatter. PAMELA was sensitive to protons and helium and their anti-partners, and was the first to measure separate spectra of protons and helium in the cosmic radiation up to energies of 10^{14} eV. However, PAMELA did not detect anti-protons or anti-helium in the cosmic radiation.

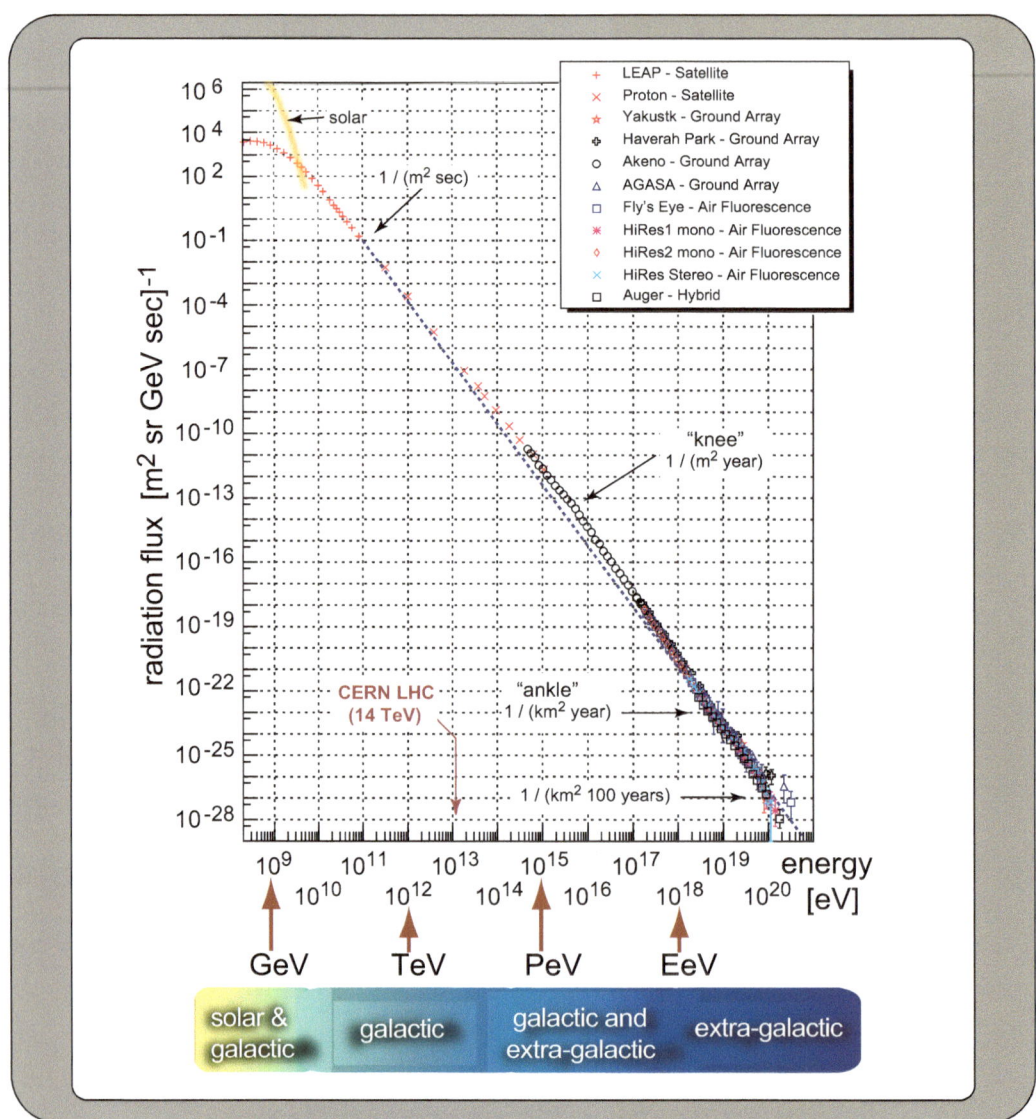

Fig. 12.12: Spectrum of cosmic rays. In this double-logarithmic representation, subtleties in the graph are difficult to discern. The energy dependence follows an $E^{-\gamma}$-law with an initial value of $\gamma = 2.6$, which slowly changes to even larger values from about 10^{15} eV onwards, corresponding to an even deeper downward slope. Moreover, the figure also indicates that, in terms of energies, there is a gradual transition from a near solar/galactic region to a potentially extragalactic region, where the sources of radiation are thought to be located. Impact frequencies are given for orientation: above 10^{11} eV about 1 impact per square meter and second, above 10^{16} eV about 1 impact per square kilometer and year, and above 10^{20} eV about 1 impact per square kilometer and 100 years. The latter value results in about 500 impacts per hour calculated for the entire surface of the Earth. [Source: whanlon@cosmic.utah.edu, adapted for this book]

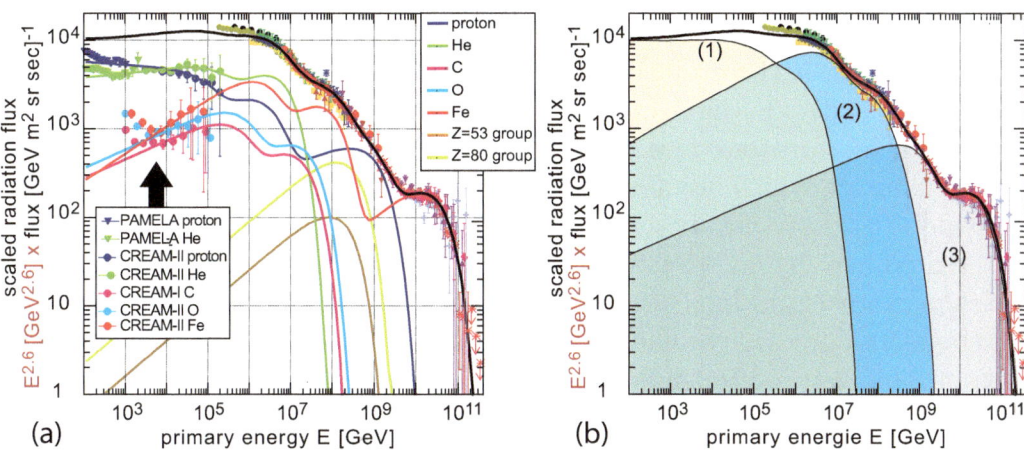

Fig. 12.13: Spectrum of cosmic rays in the range of $10^{11} - 10^{20}$ eV ($10^2 - 10^{11}$ GeV). Multiplication by $E^{2.6}$ clearly highlights the structures around 10^6 GeV ("knee") and 10^{10} GeV ("ankle"). Furthermore, data from the PAMELA and CREAM-II experiments are plotted in (a), showing particle composition up to about 10^5 GeV. They are anchor points for calculations, one of which is shown here as an example in the form of colored solid lines. These add up to the solid line in black. In (b) these calculations are divided into three regions, a theoretically well understood region (1) with the lightest particles protons and helium, a region (2) with the particles around carbon and oxygen which are also common in supernova explosions, and a region (3) which is theoretically less certain in detail and dominated by particles in the iron region. In this illustration, an extragalactic fraction below 10^{11} GeV would not be significant. [Adapted from T. K. Gaisser et al., arXiv:1303.3565v1 [astro-ph.HE] (2013)]

CREAM-II (***C**osmic **R**ay **E**nergetics **A**nd **M**ass*) was a balloon experiment launched from McMurdo Station in Antarctica in December 2005. After 28 days and a round-trip flight around the South Pole at an altitude of about 40 km, it returned with a set of amazing data about cosmic radiation. Heavy atomic nuclei such as C (carbon), N (nitrogen), O (oxygen), Ne (neon), Mg (magnesium), Si (silicon) and Fe (iron) with energies up to 10^{14} eV (10^5 GeV) could be detected in the cosmic radiation whereby the energy dependence followed the power law $E^{-(2.66\pm0.04)}$.

The additional data from PAMELA and CREAM-II are now important anchor points for theoretical models. Calculations within the framework of one of these models are shown as solid lines in Fig. 12.13 (a).

In the (b)-part of the figure these results are furthermore divided into three regions, a theoretically well understood and largely uncontroversial region (1) with the lightest and most abundant hadronic particles in the Universe like protons and helium, a region (2) with mostly carbon and oxygen nuclei, which are abundantly produced in supernova explosions, and a region (3) theoretically less secured in detail and dominated in the model presented here by nuclei in the iron/nickel region, as these are also products of supernova explosions within the Milky Way. In this representation, an extragalactic fraction at energies below 10^{11} GeV is not significant. To what extent this assumption holds can only be determined by an experimental determination of the particle species, however, the expected low counting rates makes such an endeavor a challenge. We can therefore only remark at this point that a satisfactory explanation of the cosmic radiation in this extreme energy range is still largely outstanding.

Searching for the sources

Interestingly, it is almost unavoidable that the highest-energy particles observed on Earth in the cosmic radiation are produced in the immediate vicinity of our "cosmic front door". It is the cosmic microwave background radiation, which defines a clear boundary here for nearly all particles. The faster a particle moves against the cosmic microwave background, the more this particle will "see" this radiation blue-shifted to shorter wavelengths (higher frequencies) due to the Doppler effect. This is also true for extremely relativistic particles, where their velocities change only minimally with increasing energy [10]. For instance, a proton with an energy of about $(1 - 5) \times 10^{20}$ eV sees the cosmic microwave background approaching as a hard γ-radiation, which has just enough energy to initiate the reaction

$$p + \gamma \longrightarrow p + \pi^0 \qquad \text{energy threshold in this collision} \quad 145\,\text{MeV}$$
$$p + \gamma \longrightarrow n + \pi^+ \qquad \text{energy threshold in this collision} \quad 150\,\text{MeV}$$
$$\;\;\raisebox{0pt}{$\llcorner$}\!\!\longrightarrow p + e^- + \overline{\nu}$$

The pions decay after a short path, and the proton loses close to half of

---

[10] The Doppler shifted frequency f seen by a particle with velocity $\beta = v/c$ when moving against the cosmic microwave background ($T = 2.73$ K) is calculated to: $f = f_{2.73} \cdot \sqrt{(1 + \beta)/(1 - \beta)}$.

its initial energy. After a little formula physics, one finds that the mean free path for the proton is just $60 - 100$ million light-years. The "ankle" in the spectrum of the cosmic radiation (see Fig. 12.13) may therefore have arisen from the accumulation of protons whose energies were initially greater than 10^{20} eV. This "cut-off" energy is known in the literature as "GZK cut-off", named after Greisen, Zatsepin, and Kuz'min [11]. The relativistic neutron in the above equation decays back to a proton after about a few million light-years of travel. One may therefore conclude, if protons with energies greater than a few 10^{20} eV are being observed in the cosmic-ray spectrum, the associated sources ought to be found within a volume of about 100 million light-years in radius.

A similar reasoning, though with entirely different consequences, can be developed, should the highest-energy particle components consist of atomic nuclei with masses greater than carbon or oxygen, for example. For these nuclei, a γ-energy of about $8 - 15$ MeV is already sufficient to photo-dissociate them during their intergalactic journey. In this case, a nucleon, i.e., either a proton or neutron, is knocked out of the nucleus by the high-energy photon. The significantly lower energy required compared to the pion production above causes the "cut-off" energies for these massive, thus "less fast" systems to be approximately equal to that for protons. However, the mean free path for a nuclear photo-dissociation is only about $50,000 - 100,000$ light-years, i.e., much shorter than the distances between two galaxies, and because the energy loss in photo-dissociation is less than 4%, the process is cumulative over intergalactic distances and ends with the complete destruction of the nucleus. Accumulation of heavy nuclei just below "cut-off" as would be the case for protons is therefore not expected. Hence, if the "ankle" in the spectrum of the cosmic radiation is predominantly caused by iron/nickel nuclei, as shown in Fig. 12.13, then this component can only have originated within the Milky Way and in a nearly pristine form.

Nevertheless, the localization of a cosmic-ray source remains problematic even with a precise determination of the impact direction of the particle arriving on Earth. In fact, with the exception of the neutrino, all other

---

[11] G. T. Zatsepin, V. A. Kuz'min, *Upper Limit of the Spectrum of Cosmic Rays.*, Journal of Experimental and Theoretical Physics Letters 4, 78 (1966), and
Kenneth Greisen, *End to the Cosmic-Ray Spectrum?*, Physical Review Letters 16, 748 (1966).

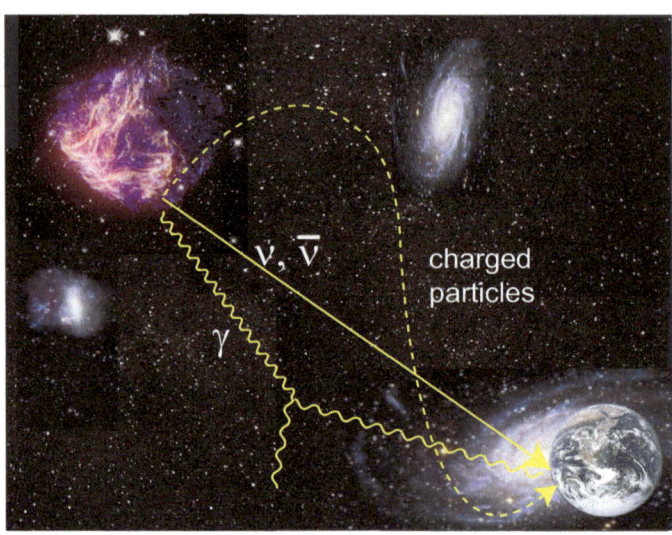

Fig. 12.14: Deflection of charged particles by galactic and intergalactic magnetic fields. Only photons and neutrinos are not affected by these fields. However, photons and, of course, charged particles can be scattered by intergalactic dust and thereby thrown out of their original path. Neutrinos, on the other hand, don't suffer any of these fates. They can travel cosmological distances unhindered.

cosmic-ray particles known so far are subject to electromagnetic interaction. Over large distances, therefore, directional changes either due to scattering or due to magnetic field deflection cannot be excluded (see Fig. 12.14). Galactic and/or intergalactic ion currents are the generators of these magnetic fields, where the field lines extend over regions far beyond the galaxies. Estimates of magnetic field strengths in intergalactic space are about $0.01 - 10$ pT (pico-Tesla) depending on the area under consideration. These field strengths are miniscule compared to the Earth's magnetic field, which is about $30\,\mu$T, but on a cosmic scale quite remarkable. A charged, highest-energy particle with an energy of 10^{20} eV would be held in such an intergalactic field on a circular orbit with a radius of $0.03 - 30$ million light-years.

Neutrinos, on the other hand, can traverse cosmic distances unimpeded and indicate the exact point of their origin. Moreover, their energies are not degraded by the "cut-off" described above. The hope is therefore that they will lift the "curtain" at $10^{20}/10^{21}$ eV and open a viewing window to new and unknown physics. Unfortunately, neutrinos also pass equally unimpeded through neutrino detectors set up by humans. — *There is no free lunch.* — The only thing that helps here is simple detector size and persistent waiting until finally one of these particles is caught every once in a while.

Cosmic tennis rackets

As energies up to 10^{20} eV and above are readily being observed in the cosmic radiation, the question arises, how these enormous energies can be produced. Clearly, sun-like stars, supernovae, neutron stars are not nearly capable of producing particles with such energies in a direct process. The limits are here at best about 10^{16} eV. However, supernovae, neutron stars and rotating Black Holes eject large amounts of matter in the form of high-energy charged particle streams, which eventually meet the quasi-thermal interstellar gas, where they experience a sudden deceleration and compression. The result is the formation of a gigantic shock front. The physics of shock fronts is well studied, yet a full description of these processes when combined with electromagnetic fields is extraordinarily complex and multi-facetted and cannot be mastered without large-scale computing facilities. Plasma physics in its full breadth comes into play here. Turbulences creating instabilities in the most extreme form, and the swirls of strong electric and magnetic fields are easily capable of either accelerating, decelerating or reflecting high-volume domains at the various impact boundaries. Large energy changes are the result. The processes are of course highly diffusive, i.e., any information about the initial direction of an object entering into this area is lost.

Despite its complexity, the process of an energy gain from a shock front in its final outcome can be easily understood by means of simple analogies from everyday life. Two examples may clarify this (see also Fig. 12.15):

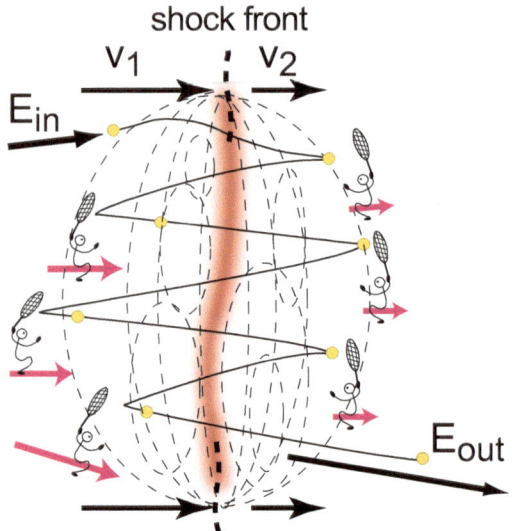

Fig. 12.15: Simplified visualization of a shock-front acceleration based on the tennis-racket effect. The shock front occurs when fast particle streams (e.g., from supernovae) hit interstellar gas and are decelerated and compressed (velocity $v_1 \gg v_2$). Turbulences and extreme magnetic field swirls are the result. A following object can gain energy if it penetrates the shock front in the direction of motion, but also if it is reflected back into the front. The process is cumulative and can end with significant energy gain ($E_{out} \gg E_{in}$). In this illustration, the tennis rackets represent the extreme fields that reflect and accelerate the charged particles. The process is highly diffusive, i.e., information about the initial direction of motion of the particle is lost.

● The tennis racket phenomenon: When serving in tennis, the tennis ball reaches speeds of up to 250 km/h (assuming tennis professionals!). Neither the hitting arm nor the tennis racket have a speed even close to this. Here, the strings of the tennis racket take the function of a shock front, and the restoring force of the strings causes the reflection and the acceleration, similar to the magnetic field in a cosmic shock front.

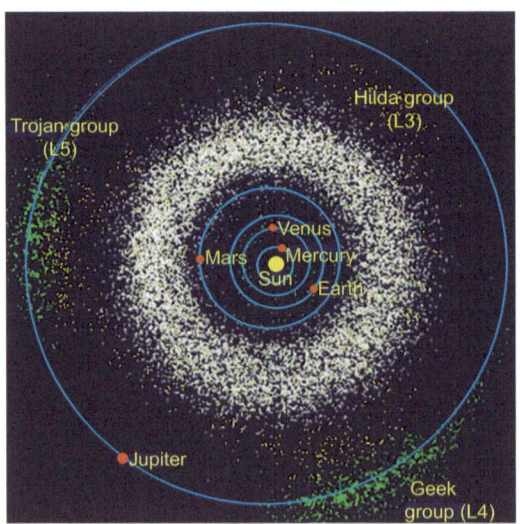

Fig. 12.16: Planetary system and asteroid belt. In the Lagrange points L3, L4 and L5 of the Jupiter-Sun system there are the so-called Trojan camps, which contain several million small asteroids, of which about 7000 are catalogued. The Cassini spacecraft used a double fly-by to Venus and thus had enough "swing" to reach effortlessly the asteroid belt located between Mars and Jupiter.

• The fly-by phenomenon: A similar, but completely contact-free situation occurs in satellite physics during interplanetary missions. A satellite approaching a celestial object (e.g., Venus, Mars, etc.) from behind, i.e., following its orbital direction of motion, is captured by the gravitational field and ejected from the gravitational field at a higher velocity in a fly-by. The faster the celestial body moves (e.g., Venus, Earth, Mars), the greater the gain in velocity. The already mentioned Cassini mission is probably the best known project in the public eye. The spacecraft completed two fly-bys to the inner planet Venus (on April 26, 1998 and June 24, 1999) and thus had enough "swing" to reach the asteroid belt between Mars and Jupiter (see Fig. 12.16). Its Huygens module eventually landed on Saturn's moon Titan in 2005.

The two processes, tennis racket and fly-by, can be calculated in closed mathematical form and are already treated in lower semesters of a physics course.

Astronomical observations, of course, help to test and further develop the mathematical description models of such cosmic shock fronts. Yet, the central question of whether cosmic shock fronts can generally be considered as the generators of the observed particle energies up to 10^{20} eV and above still remains controversial.

Supernovae and cosmic rays

The most frequent forms of shock fronts within a galaxy like the Milky Way arise as a consequence of supernova explosions when the remnants of the original star hit the interstellar gas at high velocity. Therefore, if the high- and highest-energy cosmic rays measured on Earth originate here, the particle composition of the radiation must be a reflection of the supernova nucleosynthesis with a strongly pronounced iron/nickel component typical for it, as shown e.g., in Fig. 12.13. With a frequency of about 2 supernovae per century in the Milky Way and an active lifetime of supernova shock fronts greater than 10^6 years, this would result in a nearly constant cosmic-ray flux over cosmic time scales. Moreover, since the formation processes are diffusive, the intragalactic cosmic rays would be largely isotropic. These arguments are consistent with current observations.

If the cosmic radiation created inside a galaxy is not impinging on objects like the Earth, for example, it may leave the galaxy. From the density of the radiation flux hitting Earth and assuming isotropy, it is easy to calculate the energy per unit time (equivalent to the power) outflowing from the Milky Way galaxy. The power injected by supernovae into the galaxy is on average about 6×10^{34} watt (at 2 supernovae per century with an average explosion energy of 10^{44} watt-seconds, not including neutrinos). The calculated energy outflow per unit time is about 10% of this value, so the radiative content of a galaxy can easily be maintained at a constant level over cosmic times by supernova explosions. Should our neighboring galaxies have a similar energy outflow, this radiation would be detectable with the largest Earth-based radiation detectors (however, see footnote[12]).

12.3   Gravitational waves, the new radiation

It is quite simple: If General Relativity is to be a correct theory, then the existence of gravitational waves is unavoidable. Albert Einstein had postulated this already conclusively in 1916. However, his initial derivation contained a decisive error, so that he felt compelled at the beginning of 1918, i.e., about one and a half years later, to correct this error in a somewhat more detailed, 14-page paper[13]. This publication begins with a memorable sentence:

Albert Einstein 1918: ***Die wichtige Frage, wie die Ausbreitung der Gravitationsfelder erfolgt, ist schon vor anderthalb Jahre in einer Akademiearbeit von mir behandelt worden*** [Ref.]. ***Da aber meine***

---

[12] A galaxy is in general also surrounded by intergalactic gas. If a high-energy particle flux hits this gas, a shock front is formed here as well, which throws back some fraction of the particle flux. Unfortunately, little experimental data material is available for this. A quite generally understandable theoretical treatise to this well known problem is available in the reference: Lukas Merten, Chad Bustard, Ellen G. Zweibel and Julia Becker Tjus, *The Propagation of Cosmic Rays From the Galactic Wind Termination Shock: Back to the Galaxy?*, arXiv:1803.08376v2 [astro-ph.HE], (2018) and same authors, same title in The Astrophysical Journal, 859, 63 (2018).

[13] Albert Einstein, *Über Gravitationswellen*, Sitzungsbericht, Königlich Preußische Akademie der Wissenschaften zu Berlin 1918, 154 (1918).

Darstellung des Gegenstandes nicht genügend durchsichtig und außerdem durch einen bedauerlichen Rechenfehler verunstaltet ist, muß ich hier nochmals auf die Angelegenheit zurückkommen.

Translated: ***The important question, how the propagation of the gravitational fields takes place, has been treated already one and a half year ago in an academy work by me*** [Ref.]. ***But since my exposition of the subject is not sufficiently transparent and, moreover, is marred by a regrettable arithmetical error, I must return to the matter here again***

[Ref.]:   Sitzungsbericht Königlich Preußische Akademie
der Wissenschaften zu Berlin 1916, page 648 ff

The phenomenon «*gravitational wave*» is quite intuitive and simple to understand. Since every mass is surrounded by a gravitational potential, which acts directly on its environment, a movement of the mass and with it the change of the gravitational <u>and</u> space-time conditions must also be communicated to that environment. Thereby, it is assumed that this information exchange takes place with the speed of light. From Einstein's field equations a wave equation for these time dependent fields can readily be derived with a little, though not quite easy to understand mathematics. Surprisingly, this wave equation is very similar to Maxwell's wave equation for electromagnetic fields. Gravitational waves, however, act on space-time, and since the «*gravitational*» interaction is about 36 !! orders of magnitude weaker than the «*electromagnetic*» interaction, the "feelable" effect is also smaller by a corresponding factor (see also Table 4.1). This also means that even a human being, when moving, radiates gravitational waves, and this as a direct consequence of the change of space-time caused by this movement. So far the effect is not measurable and presently merely theoretical. In fact, even until the 1960/1970s gravitational waves were generally considered unmeasurable because of their tiny interaction strength.

The first, albeit indirect, evidence of energy radiation by gravitational waves was obtained in 1974 by the astronomers Russell A. Hulse and Joseph H. Taylor [14], who conducted an impressive 25 year long and extremely precise observation of the orbit of the millisecond pulsar PSR1913+16, which is located at a distance of 20,870 light-years and has a rotation period of about 59 ms. The pulsar forms a binary star system with an invisible neutron star. Both objects have an almost identical mass of about 1.4 solar masses, which is a typical value for neutron stars. They are in

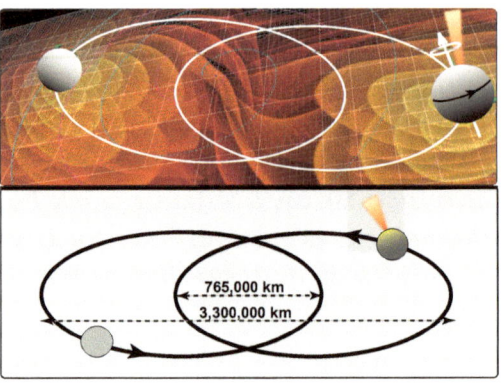

Fig. 12.17: Illustration of the pulsar binary system PSR1913+16, in which two neutron stars (pulsar on the right) rotate around each other in eccentric orbits, emitting gravitational waves. The current orbital period of 7.75 hours changes by 76.5 μs per year.

an eccentric elliptical orbit around a common center-of-mass with an orbital period of 7.75 hours. Their distance varies from 765,000 km (minimum) to 3.34 million km (maximum) (see Fig. 12.17). Hulse and Taylor could show by their measurements that the orbital period of 7.75 hours decreases successively due to gravitational waves radiation. The present exact value of this change is 76.5 μs per year, and the ratio of this measured value and that calculated by General Relativity is 0.997 ± 0.002 — an astounding result in every respect [15]. Further, due to the constant energy output by gravitational waves, the major semi-axes of the respective orbits shrink by about 3.5 m per year. The radiated power of this system is about 7.35×10^{21} kW. This corresponds to about 2% of the light power emitted by the Sun. With these data, the binary star system has about 300 million years left before it collapses to a Black Hole.

The discovery of Hulse and Taylor was undoubtedly a highlight in the

---

[14] *"For the discovery of a new type of pulsar, which opens up new possibilities for the study of gravity."* Russell A. Hulse and Joseph H. Taylor, Jr., both of Princeton University, USA, received the 1993 Nobel Prize in physics.

[15] Joel M. Weisberg, David J. Nice, Joseph H. Taylor *Timing Measurements of the Relativistic Binary Pulsar PSR B1913+16*, Astrophysical Journal 722, 1030 (2010) and same authors, same title arXiv:1011.0718 [astro-ph.GA].

research on gravitation. Nevertheless, still today surprisingly little is known about this interaction and much is still subject to untested and at times speculative theoretical models. Promising laser technologies have eventually triggered the development and construction of gravitational wave detectors with the aim to measure gravitational waves in a direct way. The two largest and by now already active projects are the LIGO Observatory (LIGO=**L**aser **I**nterferometer **G**ravitational wave **O**bservatory) in the US and the VIRGO project initiated by France and Italy on the site of the European Gravitational Observatory (EGO) in Cascina near the city of Pisa. The techniques were designed to capture the relative length changes of $\Delta l/l \leq 10^{-21}$ as these are the expected effects when gravitational waves created by cosmic events pass through the detector. On a scale of 1 meter, this specification corresponds to about one millionth of a proton diameter (or one 100 billionth of the diameter of a hydrogen atom). LIGO and VIRGO have since undergone significant upgrades to increase their sensitivities and are by now known in the literature as "Advanced-LIGO" and "Advanced-VIRGO" (see Fig. 12.18).

The Advanced-LIGO Observatory consists of two gigantic and widely separated Michelson interferometers, one at Hanford in Washington state and the other about 3700 km away at Livingston in Louisiana. The two instruments are synchronized to correlate possible signals from gravitational waves once they pass through both detectors, thereby allowing a suppression of locally induced spurious components from different sources. After a 10-year construction period followed by about 5 years of testing, LIGO began taking its first data in 2007.

At about the same time, the VIRGO interferometer went into the construction phase, which, after an upgrade pause, began scientific operation as "Advanced-VIRGO" on August 1, 2017. With these three instruments, it is now possible for the first time to accurately locate the source of radiation, because unlike electromagnetic waves (i.e., light), gravitational waves have a different formation characteristic (see Fig. 12.19). To locate a source, it is necessary to measure the time differences between gravitational waves arriving at three detectors minimum (and located as far apart as possible).

The principle of operation of a Michelson interferometer is shown in Fig. 12.20 using the relevant information from the LIGO experiment. The

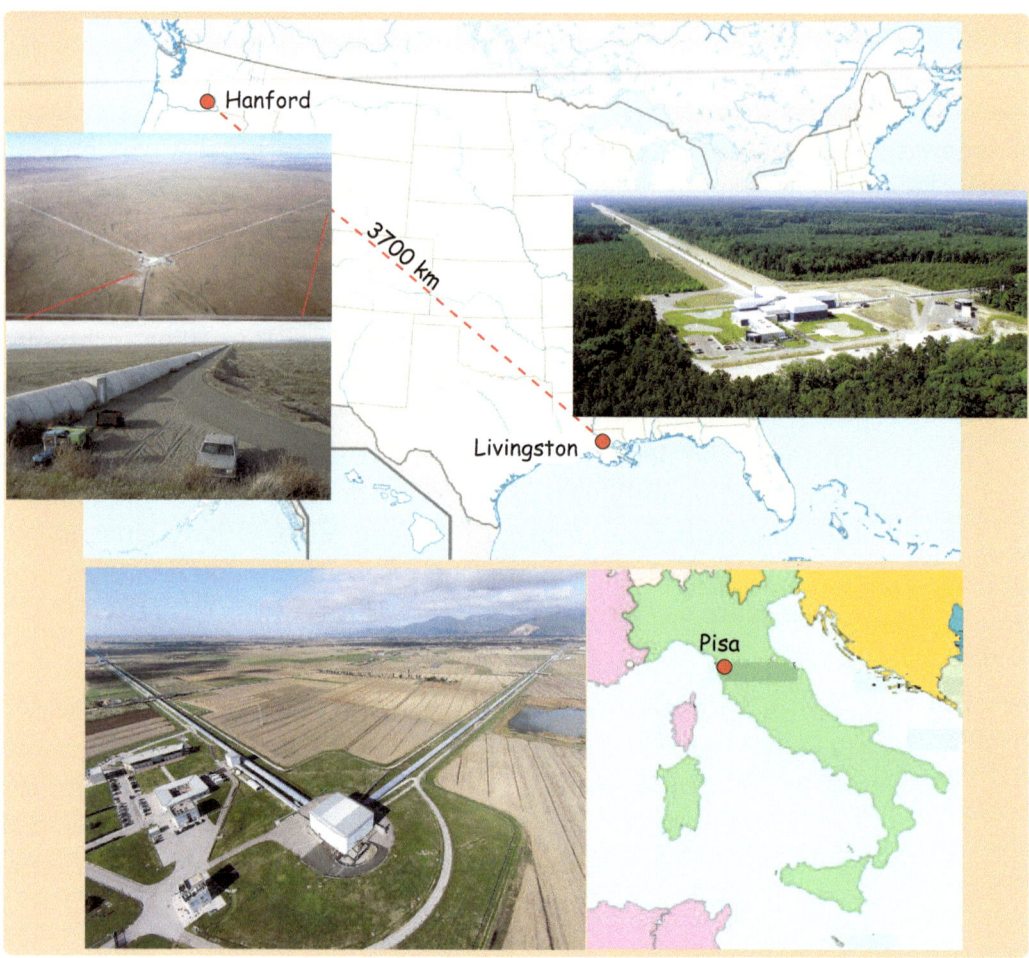

Fig. 12.18: The LIGO and VIRGO gravitational waves detectors in the US and Italy. The aerial photographs show the extents of the interferometer arrays. The arm lengths of LIGO are 2 km at Hanford and 4 km at Livingston, and those of VIRGO are 3 km.

beam power of a 180 W frequency-stabilized laser is first increased to about 5 kW in a "power recycling" mirror. Via a beam splitter, the beam is directed into two perpendicular arms containing the respective Fabry-Pérot resonators. Each resonator consists of two mirrors, one of which (the entrance mirror) is partially transparent. The light passes through the 4 km long path in the resonator 280 times before it hits the beam splitter again through the partially transparent mirror and from there reaches the light-sensitive photodiode. The multiple reflection technique increases the effective travel length of the light to $4 \times 280 = 1120$ km, resulting in a

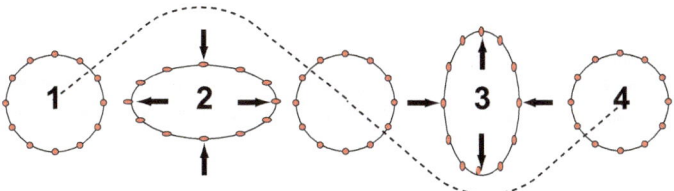

Fig. 12.19: The principle of a gravitational wave: Within a whole wavelength (1 - 4) a longitudinal (2) and a transverse (3) contraction occur at intervals of half the wavelength each. At a typical frequency of about 300 Hz, the wavelength extends over a range of 1000 km.

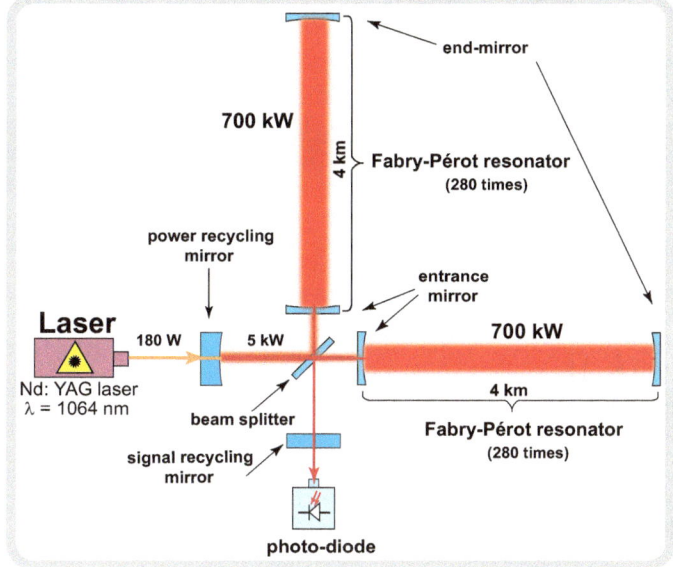

Fig. 12.20: Operational mode of the LIGO Michelson interferometer.

significant increase in the sensitivity of the instrument. The entire arrangement is, of course, located in an ultra-high vacuum.

If a gravitational wave crosses the interferometer, the relative lengths of the two arms change. In the wave crest of the gravitational wave the length is shortened, in the wave trough it is lengthened. Because of the rectangular arrangement, the effect is different in the two arms. This causes a tiny frequency shift, which manifests itself as a phase shift of the two partial waves of the laser light and, when superimposed, changes the interference profile in the light detector. The zero position is set so that ideally the

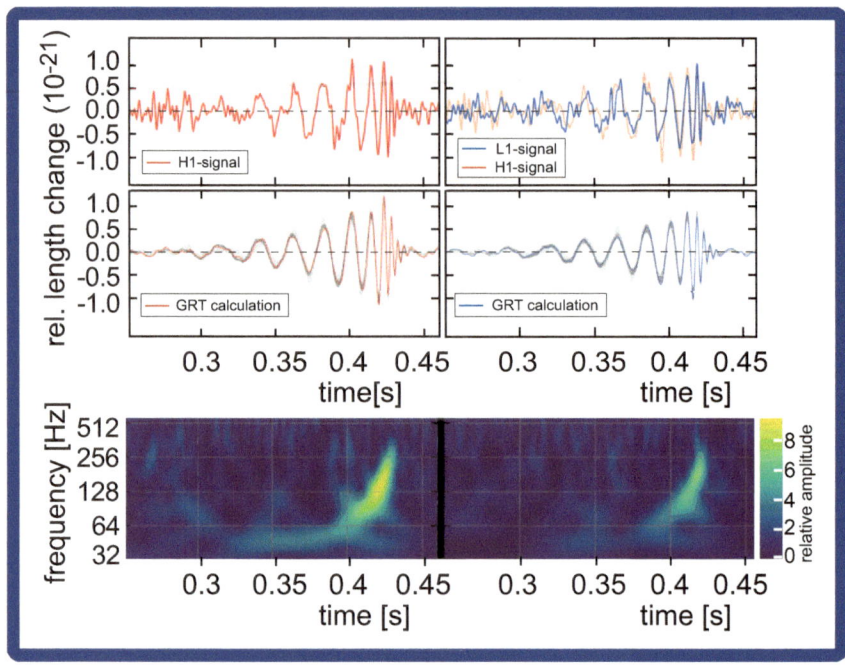

Fig. 12.21: Gravitational wave signal from the collapse of two Black Holes measured by the "Advanced-LIGO" detector on September 14, 2015. The above image shows the oscillation characteristics caused by the orbital frequency of the two Black Holes, in red the signal in the Hanford detector (H1-signal) and in blue the signal in the Livingston detector (L1-signal) — below are shown calculations according to the General Relativity Theory (GRT). The bottom panel shows the frequency analysis of the signal with the maximum at about 150 Hz. The entire time scale of the Black Hole merging event extends over barely 0.2 s

laser light exactly cancels at the location of the light detector. Frequency and intensity of the gravitational wave are thus directly transferred to the current of the photo diode.

On September 14, 2015 — almost 100 years after Einstein's prediction — the LIGO collaboration achieved the first direct detection of gravitational waves. The measured signal (see Fig. 12.21) came from two Black Holes rotating around each other, moving faster and faster towards each other, until they finally collided and coalesced. One Black Hole of about 36 and one of about 29 solar masses had merged, and a Black Hole of about 62 solar masses remained. The difference of 3.0 ± 0.5 !! solar masses was thereby completely converted into energy of gravitational waves and sent on its journey through the Universe. The whole event took place in less than a second and happened at a distance of about 1.3 ± 0.5 billion light-years. Arriving at Earth, the gravitational waves created a relative space distortion on the order of $\Delta l/l \approx 10^{-21}$ with an oscillation frequency corresponding to the orbital frequency of the two Black Holes during their merging process [16].

On August 14, 2017, at 12:30:43 CET, both LIGO detectors and the VIRGO detector observed the signal GW170814 (**G**ravitational **W**ave-14-08-2017), which was also generated by two coalescing Black Holes — significant also because Advanced-VIRGO had just begun observations on August 1st.

On August 17, 2017, electromagnetic radiation and gravitational waves from an event in the galaxy NGC 4993, 130 million light-years away, were registered simultaneously for the first time. The analysis pointed to the coalescence of two neutron stars.

In October 2020, the LIGO and VIRGO collaborations published an updated gravitational wave catalog with 50 entries now, 46 of those are events associated with merging Black Holes [17]. Of course, the catalog gets constantly updated over time.

Outlook

The successful detection of gravitational waves opens a completely new, unknown and exciting chapter in cosmic research. The fact that mergers of neutron stars and Black Holes had been observed within a very short time after the turning on of the gravitational wave detectors shows that such events are not rare in the Universe. This holds a considerable discovery potential, which is by no means only limited to cosmological questions. Some remarks on this are therefore given here:

▶   Gravitational waves emitted by Black Holes during collision or merging give an immediate insight into the whole spectrum of oddities surrounding Black Holes. Such insight cannot be obtained by other types of radiation.

---

[16] In 2017, Rainer Weiss, Mass. Inst. of Technology (MIT), Barry Barish, Calif. Inst. of Technology (Caltech) and Kip Thorne (Caltech) received the Nobel Prize in Physics for this work.

[17] URL: www.ligo.org/science/Publication-O3aCatalog/

The main question is, what physical laws describe these objects and to what extent may they change our current view of the world.

▶   Primordial gravitational waves carry information about the earliest phases of the Universe and allow an unobstructed view up to the inflationary phase. Neither neutrinos nor photons allow such a look back into the past.

▶   The questions of how time and space arise and how these quantities are subject to quantization are also directly touched upon by the study of gravitational waves. A quantum gravity theory still awaits discovery and experimental verification.

▶   Through the merging of Black Holes sometimes immense masses are converted into gravitational wave energy bypassing the baryon number and lepton number conservation, which have so far been firmly established in particle physics and been respected by all other interactions to utmost precision. Elementary particle physics will have to deal with this new phenomenon in the context of a comprehensive quantum theory.

▶   The internal matter state of neutron stars and pulsars remains poorly understood. To what densities can matter be compressed and what new forms of states are formed here? Do the superconducting forms of baryonic matter predicted by theoretical models exist? Gravitational waves have the potential to provide information about this.

▶   Finally, how do gravitational waves interact with dark matter and dark energy, which are, after all, the largest components in the Universe?

12.4   Three detectors

There are currently a total of 41 different and active experimental projects investigating and unraveling cosmic rays, of which 24 are Earth-based experiments, 10 are satellite experiments — the oldest of which is Voyager-1 with now over 43 years in space and about 22 light-hours away at the end of 2020 — and 7 are balloon experiments. Each one pursues specific questions. We will here briefly describe only the three largest and — because of their size — most spectacular detectors. More detailed descriptions of these detectors, their modes of operation, and the scientific goals of the projects can be found on the respective Internet platforms.

The "IceCube" detector

IceCube is a high-energy neutrino observatory in Antarctica and part of the local Amundsen-Scott South Pole Station. The station lies on an ice sheet 2835 m above sea level. The Observatory operates the IceCube detector within an international collaboration. IceCube is specifically designed to detect high- and ultrahigh-energy neutrinos in cosmic rays. Since neutrinos are not deflected by cosmic magnetic fields and also penetrate matter almost unimpeded, cosmic sources can be located from their direction of incidence. Together with astronomical observations carried out at other locations on Earth, attempts are being made to decipher the creation mechanism of highest-energy neutrinos.

The detector material of IceCube is the pristine glacier ice at the South Pole. This ice has a particularly advantageous property, which was ultimately the reason for installing a high-energy neutrino detector at this inhospitable location. It is characterized by exceptional purity and is therefore of unique transparency to the Čerenkov radiation generated by charged particles when these are produced, for example, as a result of a neutrino interaction. A total of 5160 individual light-sensitive detectors (photomultipliers) suspended on 86 cable strands are embedded in the ice down to a depth of 2450 m. These make up the IceCube detector with its $1 \, \text{km}^3$ volume and give it its hexagonal structure. The individual detectors are strung along the cable strands like a string of pearls. By drilling a hole into the ice using hot water, each of these strands was let down into the water, and upon completion the hole was left to freeze over again. (For Čerenkov radiation, see footnote[18].)

---

[18] Čerenkov radiation arises, when a charged particle moves in a medium with superluminal velocity. This is possible because in a medium the speed of light c_M is reduced by a factor n compared to its vacuum value c_V, where n is the refractive index of the medium, i.e., $c_M = c_V/n$. The radiation is emitted in a cone-shape along the particle path, similar to the Mach cone observed in acoustics after breaking the sound barrier. Ice and water have refractive indices $n = 1.31$ and $n = 1.33$, respectively. The difference is due to the difference in density. *"For the discovery and interpretation of the Čerenkov effect"*, Pavel A. Čerenkov, Ilya M. Frank and Igor J. Tamm together received the Nobel Prize in Physics in 1958.

In addition to the neutrino detector in the ice, there is a $1\,\mathrm{km}^2$ area above IceCube (IceTop) instrumented with 162 water/ice Čerenkov detectors, which measure air shower profiles generated by cosmic rays in order to correlate these signals with those from the IceCube detector.

The photodetectors represent a special development. They have a $550\,\mathrm{cm}^2$ photo- sensitive half-spherical cathode area and are housed in a pressure-resistant glass enclosure along with the digital readout and communications electronics. Prior to installation, each glass enclosure was tested to an external pressure of about 650 atmospheres (about 6500 meters water-equivalent).

An impressive and detailed description of all technical aspects of instrumenting the IceCube detector is given in reference [19].

Figure 12.22 gives an overview of the area at the South Pole and the size of the IceCube detector above and below the ice surface.

Neutrinos in IceCube

The detection of neutrinos always goes via neutrino induced reactions which produce charged particles. This is of course also true for IceCube, where the charged particles, if they have enough energy, produce Čerenkov radiation in the ice. The higher the primary energy, the more particles are produced and the more intense the Čerenkov radiation. However, an identification of primary cosmic neutrinos, e.g., those originating from a cosmic source, is by no means trivial. This is illustrated by Fig. 12.23, which shows various scenarios of cosmic-ray reactions with the Earth and/or its atmosphere. In most cases the detection of a muon signifies that it was accompanied by a neutrino at its creation:

▶ In ① a high energy cosmic-ray particle hits the atmosphere and produces a typical air shower. In the air shower, muons and muon neutrinos (e.g., due to the decay of pions) are produced, typically in significant numbers. In

[19] IceCube Collaboration: M. G. Aartsen et al., *The IceCube Neutrino Observatory: Instrumentation and Online Systems*, arXiv:1612.05093v2 [astro-ph.IM], (2017) and same authors, same title, Journal of Instrumentation 12, P03012 (2017).

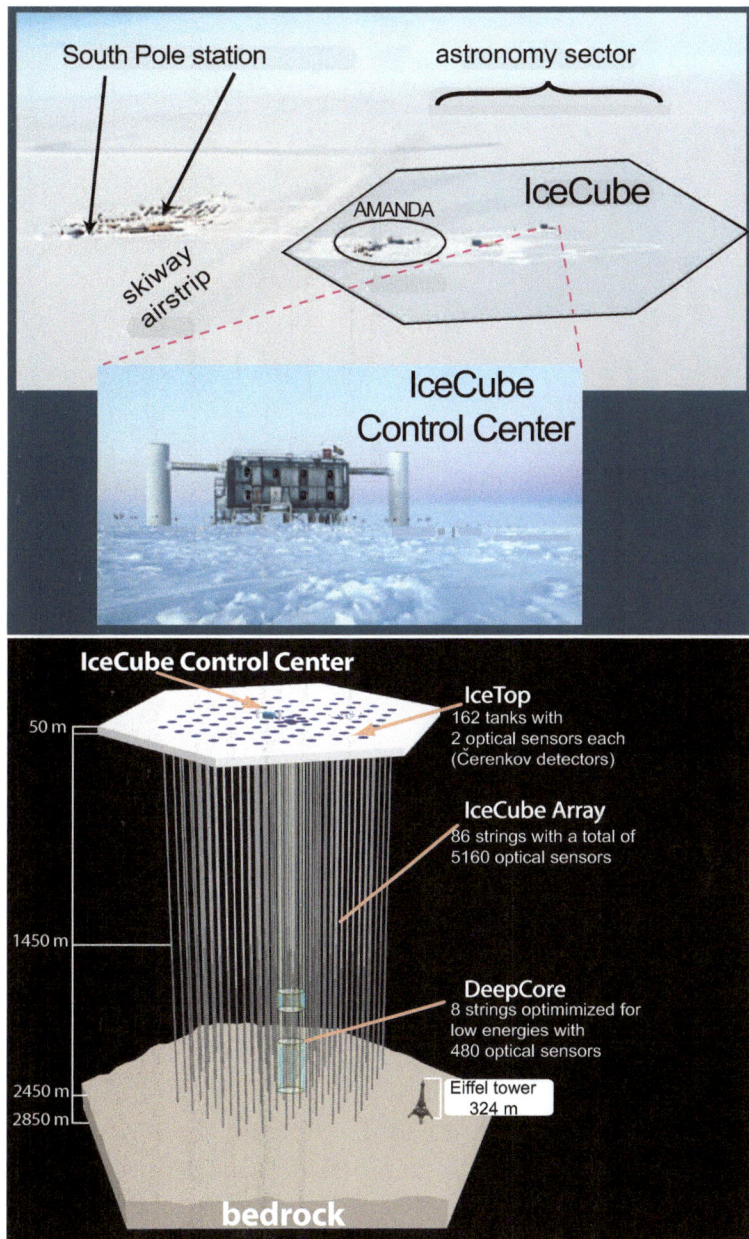

Fig. 12.22: Top: The Amundsen-Scott South Pole Station with the Antarctic Muon And Neutrino Detector Array (AMANDA) and the IceCube detector. AMANDA was a precursor to IceCube and was discontinued in 2005 after 9 years of operation.
Bottom: The interior of IceCube with the cable strands and photodetectors strung along them.

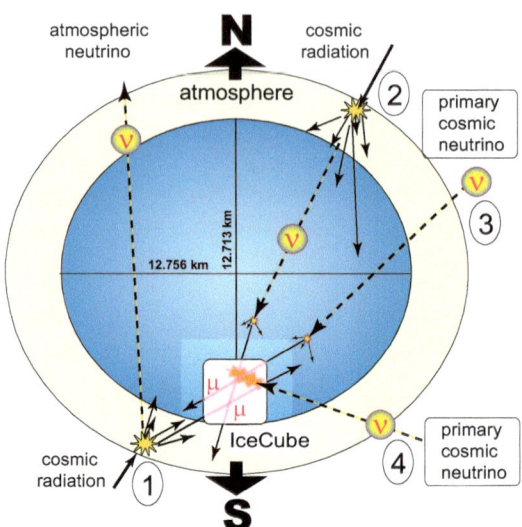

Fig. 12.23: Neutrino signatures in the IceCube detector at the South Pole. In the example reactions (1) and (2), "atmospheric neutrinos" are produced as secondary particles in the atmosphere, in (3) and (4) the reactions are initiated by primary "cosmic neutrinos" (see text).

the figure, a neutrino leaves the Earth while the muon enters the IceCube detector and gets registered. The identification of the muon is easy, since its energy loss is small. It can completely penetrate the detector.

▶ The process in (2) is also initiated by an interaction with the atmosphere with the production of a neutrino. The neutrino penetrates the Earth and in a second interaction process a muon is created, which reaches the detector.

▶ In the process (3), a primary cosmic neutrino initiates a reaction within the Earth, producing a muon.

▶ Process (4) is also triggered by a primary cosmic neutrino, but now within the IceCube detector. This reaction is particularly easy to identify because here the particle tracks are produced within the detector.

In Fig. 12.24 an event identified as a neutrino event is shown. The energy deposited in the photo-detectors as Čerenkov radiation is symbolized by the size of each "blob", and the color coding indicates the time information in a time window of about $40\,\mu$s. In the image, the neutrino enters the detector from the lower right and induces a particle shower that propagates to the upper left. A center-of-mass analysis of the signal intensities together with the time information allows an exact determination (accuracy about $0.5° - 1°$) of the direction of incidence and thus makes a localization of the neutrino source possible.

IceCube registers about 2500 events per second! Of these, muons are by far the most frequent particles. They are copiously produced as secondary particles in the air showers generated by cosmic particles. About one event in a million is identified as a high-energy neutrino event, which is also produced in the Earth's atmosphere and is therefore labeled "atmospheric neutrino." Only about $20 - 30$ neutrino events per year are of cosmic origin.

In 2017, an assignment of a cosmic neutrino to a cosmic source was achieved for the first time. The direction of the detected cosmic neutrino (event:

Fig. 12.24:   Neutrino event in the IceCube detector.   The size of each "blob" symbolizes the measured energy in the detectors and the color coding gives the time information. In this case, the particle shower starts at the bottom right of the image and propagates to the top left; "top" here means toward the surface.

[Source: https://icecube.wisc.edu/gallery]

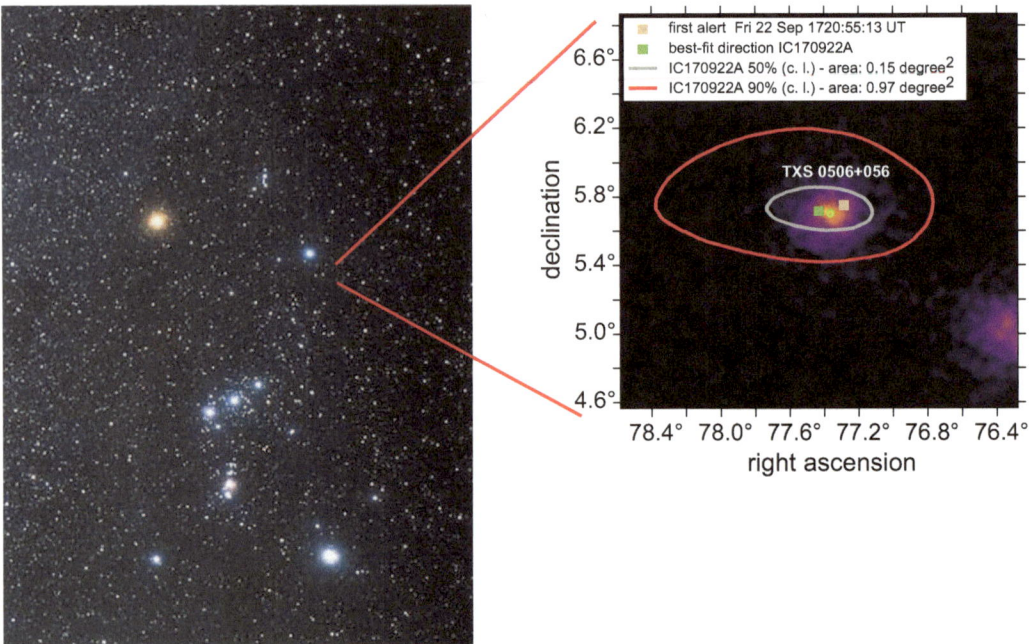

Fig. 12.25: In 2017, the IceCube neutrino event IC170922A could be assigned to the blazar "TXS 0506+056" in the constellation Orion (not visible in the image), which is 5.7 billion light-years away. The blazar developed a particularly high activity at this time. The figure shows the neutrino direction preliminary determined by IceCube and the subsequent best-fit with the 50% and 90% confidence intervals (c. l.). The right image is from the Fermi Large Area Space Telescope.

IceCube-170922A) pointed to a position in the constellation Orion. As usual in such cases, an ALERT-signal is immediately transmitted by the IceCube collaboration to all telescopes on Earth with a preliminary position indication. In this case, the cosmic source was identified as the blazar "TXS 0506+056", 5.7 billion light-years away, which is known to have its relativistic jet pointed directly at Earth. Telescope data showed that the blazar had entered a stage of particularly high activity at that time. In Fig. 12.25 this event and its tracing to the blazar "TXS 0506+056" is shown [20].

---

[20] The IceCube, Fermi-LAT, MAGIC, AGILE, ASAS-SN, HAWC, H.E.S.S, INTEGRAL, Kanata, Kiso, Kapteyn, Liverpool telescope, Subaru, Swift/NuSTAR, VERITAS, and VLA/17B-403 teams, *Multi-messenger observations of a flaring blazar coincident with high-energy neutrino IceCube-170922A*, arXiv:1807.08816v1 [astro-ph.HE] (2018).

Of the many millions of neutrinos, which the blazar had sent on their way towards the Antarctic ice at that time, this one neutrino eventually got stuck in the IceCube detector, though a re-analysis of older data revealed another handful events pointing in the same direction.

The future of IceCube: IceCube has a considerable number of scientific success stories to its credit. This makes it obvious to consider for the future a further upgrade of the detector to a $10\,\mathrm{km}^3$ sized successor "IceCube-Gen2". A wealth of questions concerning particles and particle species, their energies and properties, concerning new and unknown physics and those touching upon the mechanisms of energy generation in the Universe are expected to be brought closer to light. The goal is to improve the directional precision by a factor of 5 and in parallel to increase the efficiency for differentiating between cosmic and atmospheric neutrino signals. To achieve this, radio antennas will have to be inserted into the surface ice above IceCube, presently proposed to be placed at a depth of about $100\,\mathrm{m}$. These detect the radio emission (known as Askaryan radiation at around $100\,\mathrm{MHz}$ [21]) from a particle shower as it evolves inside the ice. This will help to improve the accuracy of determining the point of the cosmic source. In addition, a detector area on the surface-ice on top of IceCube will identify muons created in atmospheric shower events. Time correlated neutrino signals inside the IceCube detector can then be unambiguously classified as being caused by "atmospheric neutrinos". Since genuine "cosmic neutrinos" do not exhibit such a shower signature, this differentiation will lead to a significant purification of the cosmic neutrino event sample (see Fig. 12.23).

By the enlargement to approx. $10\,\mathrm{km}^3$ also the data rate will increase by a factor of 10, which opens up the possibility of advancing into energy regions beyond 10^{20} eV. The completion of IceCube-Gen2 is currently targeted for year 2033 [22].

---

[21] Askaryan radiation is emitted as part of the Čerenkov radiation, whenever an unshielded charged cloud of free electrons is created and propagates over some distance inside the medium. This usually happens when the primary particle is of highest energy. The frequency of the radiation lies in the range of a few $100\,\mathrm{MHz}$. The radiation got its name from the Armenian physicist Gurgen A. Askar'Yan (1928 – 1997), who predicted this radiation through a theoretical calculation in 1962 (G. A. Askar'yan, Zh. Eksp. Teor. Fiz. 42, 1360 (1962) [Soviet Physics JETP 15, 943 (1962)]).

[22] The IceCube-Gen2 Collaboration, *IceCube-Gen2: The Window to the Extreme Universe*, arXiv:2008.04323v1 [astro-ph.HE] (2020).

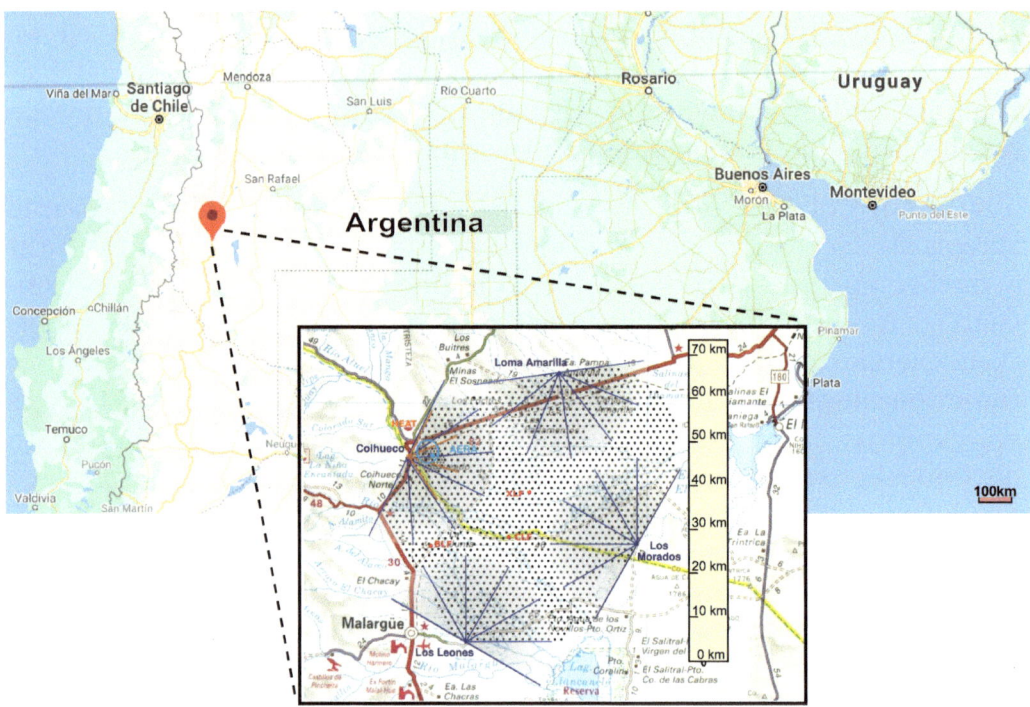

Fig. 12.26: Geographical location of the Pierre-Auger Observatory in Argentina. The lower part shows the area with the Čerenkov water tanks (marked by dots) as well as the coverage area (shaded gray) of the four fluorescence telescopes located at the periphery. These register the fluorescence light emitted by atmospheric nitrogen in the particle shower — but only possible during starry and moonless nights (see also Fig. 12.27).

The Pierre-Auger Observatory

The Pierre-Auger Observatory in the Argentinean Pampas currently constitutes the largest detector for cosmic rays. It covers an area of $3000\,\mathrm{km}^2$ (Fig. 12.26). The entire atmosphere above this area serves as detector material. This makes its active mass more than 30 times that of IceCube.

"Pierre-Auger" is an air shower detector designed to measure cosmic particle energies beyond about 10^{18} eV. It is also designed to determine the direction of the incident primary particle with high precision. For this purpose, the area in the Pampas is covered with 1660 water-Čerenkov detectors spaced $1.5\,\mathrm{km}$ apart. Each detector-unit consists of a container with about $12\,\mathrm{m}^3$ of ultra-pure water, where three photo- multipliers are embedded to register the Čerenkov radiation of the charged particles traversing the water. The

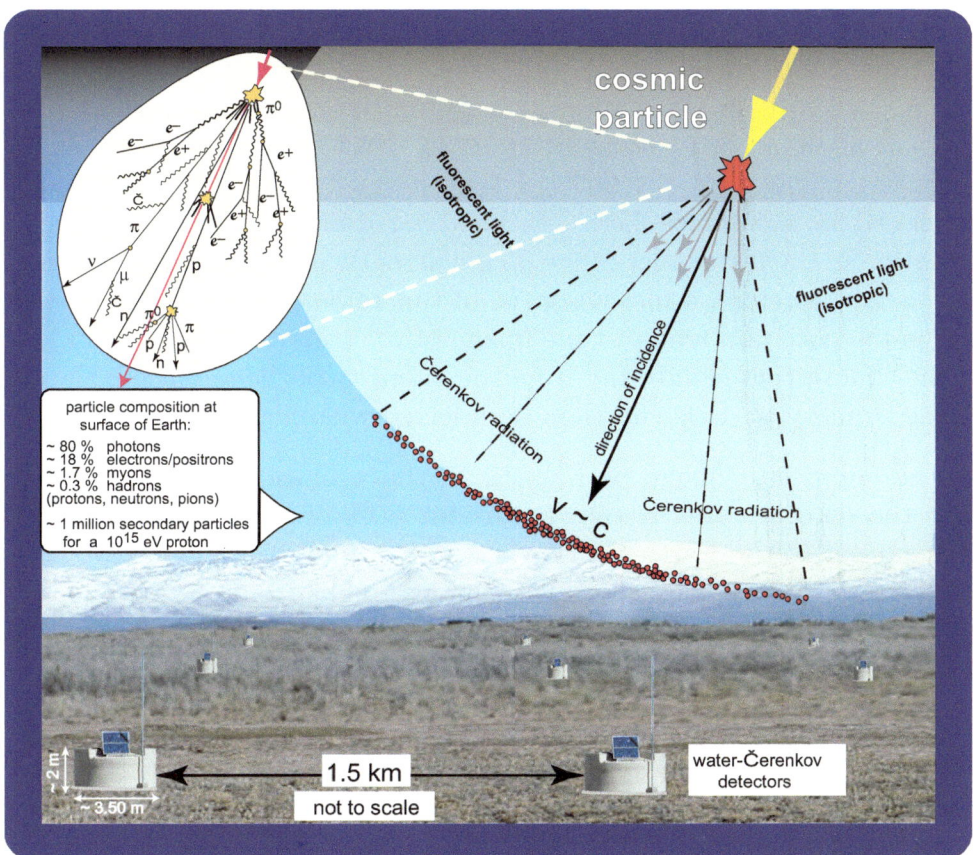

Fig. 12.27: Composite image of the Pierre-Auger Observatory in the Argentinean Pampas (the Andes mountain range in the background). In the image is inserted the shape of a particle shower that originates in the upper atmosphere. In the foreground are the tanks with the water Čerenkov detectors that register the Čerenkov radiation from secondary particles whose velocities exceed the speed of light in the medium water. In each of these, three light detectors are inserted. A transmitter-unit sends all information to a central station. Sunlight collectors generate the necessary electrical energy. The containers are about 1.5 km apart (not drawn to scale here) and cover an area of about $3000 \, \text{km}^2$.
A particle shower leaves a specific intensity and time structure in the signals. This allows the angle of incidence of the primary particle to be reconstructed with an accuracy of less than one angular degree. Atmospheric nitrogen emits fluorescent light in the ultraviolet wavelength range as a result of the interaction with the particle shower, which is registered by fluorescence telescopes installed at the periphery (see also Fig. 12.26).
[Source: Image created from templates on the "Pierre-Auger" Internet platform.]

ultra-pure water is produced on-site. A transmitter-unit transmits the information to a central station, and a sunlight collector attached to each of the containers provides all necessary electrical energy. At four corners of the "Pierre-Auger" site, fluorescent light telescopes are installed in specially constructed buildings. Inside, 6 telescopes are lined up, each about $13\,\text{m}^2$ in size. With them, the shower evolution in the atmosphere can be followed over an angle of $180°$ (horizontal) $\times$ $30°$ (vertical) — but only on starry, moonless nights. About 90% of the energy contained in the shower is deposited in the atmosphere. In the course of this process, fluorescence radiation in the ultraviolet range is concurrently emitted by the atmospheric nitrogen. Therefore, a measurement of this radiation allows the creation of a true-to-space and true-to-time image of the shower profile. From the size of the "fluorescent cloud" and from the intensity of the radiation, the energy of the primary particle can then be inferred with high precision[23].

In the composite image of Fig. 12.27 the evolution of a shower profile over the detector area is shown. About 80% of the particle components transported to the Earth's surface consist of photons, about 18% are electrons and positrons, and the rest are muons and hadrons. By then more than 90% of the primary energy has already been absorbed in the atmosphere. The wave front containing the highly relativistic particles that eventually enters the water tanks has a thickness of several meters.

The Fig. 12.28 shows a real event generated by a cosmic particle with a primary energy of 1.25×10^{19} eV. It produces a particle shower that extends over nearly $30\,\text{km}^2$ within the "Pierre-Auger" area. The intensity of the light signal received by one of the fluorescence telescopes emitted by the atmospheric nitrogen is shown along the impact direction.

The future of "Pierre-Auger": The Pierre-Auger Observatory is currently undergoing an upgrade to "AugerPrime." Here, the water Čerenkov containers will be equipped with additional external detectors (scintillation detectors) and also with classical radio-receiving antennas. The scintillators register the high energy photons as well as electrons/positrons, and the radio antennas receive the Askaryan radiation (see footnote page 301), which is emitted from extended air showers in the few 100 MHz range. The measurement of both components allows a more precise determination of

---

[23] A very much enlightening article on the extraordinary technological achievement in instrumenting the Pierre-Auger Observatory can be found in: The Pierre-Auger Collaboration, *The Fluorescence Detector of the Pierre Auger Observatory*, arXiv:0907.4282v1 [astro-ph.IM], (2009).

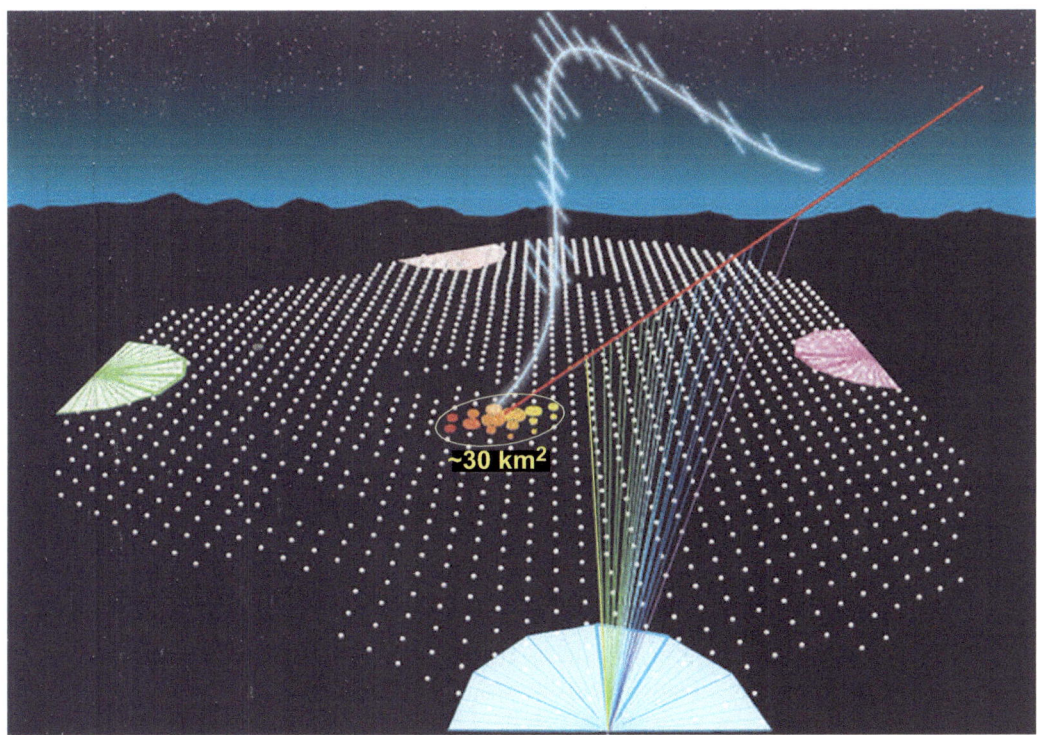

Fig. 12.28: Signature of an impact of a cosmic-ray particle into the atmosphere and the subsequent shower profile development. The particle shower propagates on the ground over an area of about 30 km². The atmospheric nitrogen light signal received by one of the fluorescence telescopes is shown along the impact path (in blue). The time profile of this signal is characterized by the color transition from violet-blue to green-orange. [Source: Pierre Auger Observatory internet platform.]

the energy and the angle of impact of the primary particle, and even more importantly, a determination of the particle mass. In a second project, the fluorescence detectors will be retrofitted in such a way that the light from the moon during night-time hours no longer requires them to be switched off. And finally, in the "AMIGA" (**A**uger **M**uons and **I**nfill for the **G**round **A**rray) sub-project, additional water-Čerenkov detectors will be installed in a selected 20 km² area at only 750 m intervals, and large-area muon detectors will be located at a depth of a few meters below them. This will make it possible to lower in this area the energy detection threshold of the primary cosmic-ray particles to about 10^{17} eV and to study in more detail the possible transition region from the galactic to the extragalactic cosmic radiation (see also Figs. 12.12 and 12.13).

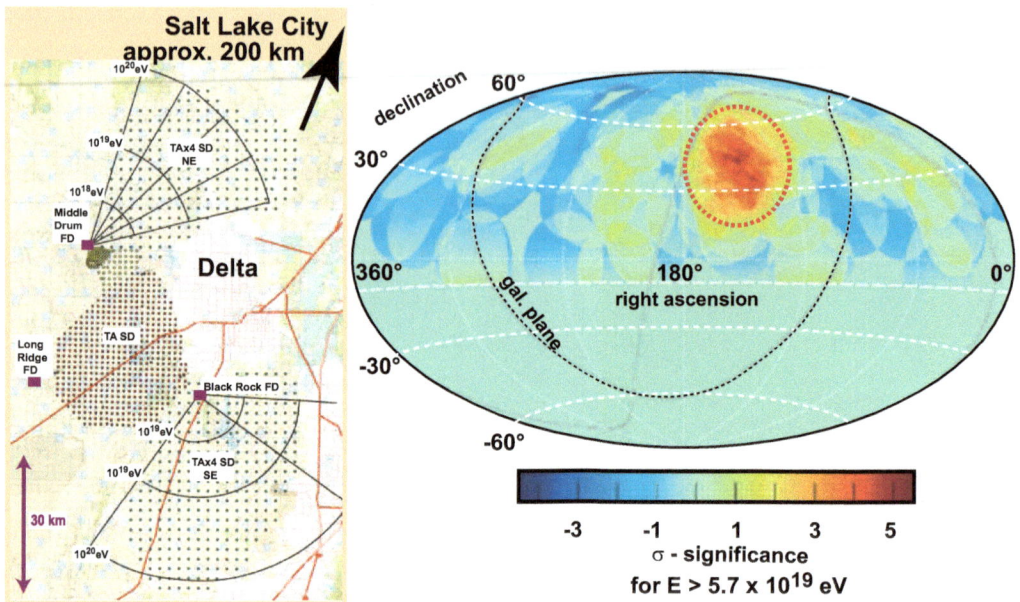

Fig. 12.29: The image on the left shows the areal layout of the Telescope Array (TA) detector and its successor TAx4 near the city of Delta, which is about 200 km away from Salt Lake City in the US state of Utah. One of the major goals is the exploration of highest-energy cosmic-ray sources in the viewable region of the Northern Hemisphere. (Abbreviations in image: SD = surface detector, FD = fluorescence detector stations, NE, SE = northeast, southeast) — The image to the right shows the significance map created by the TA collaboration for cosmic rays beyond an energy of 5.7×10^{19} eV (negative value for energies $< 5.7 \times 10^{19}$ eV). It indicates a source, which is currently receiving special attention.

The Telescope Array (TA)

The equivalent of the Pierre-Auger Observatory in the Northern Hemisphere is the Telescope Array (TA) Observatory located near Salt Lake City in the state of Utah. It uses a technology almost identical to that of the Pierre-Auger Observatory with a similar areal detector layout, which presently covers about 700 km^2. This detector is also in an upgrade stage to "TAx4" which, when completed, will cover a detector area of 2800 km^2, i.e., 4 times the present size.

The TA detector is designed for highest-energy cosmic rays beyond 10^{18} eV. It will focus as well on the transitional region between the galactic and the extragalactic components of cosmic rays, though in a somewhat higher energy regime. Over the years since its commissioning in 2008, it has been able to locate an unusually strong cosmic-ray source with energies greater than 5.7×10^{19} eV in the northern hemisphere field-of-view (see Fig. 12.29). This energy corresponds to the "GZK cutoff" energy for protons (see page 281) and suggests a strong and much higher-energy primary source behind this GZK curtain and thus beyond about 100 million light-years. In the coming years, the TAx4 detector will therefore further explore this source and pursue the question: Which cosmic object is hidden behind this curtain and which physical processes come into play here?

12.5   Something else worth knowing

Production of carbon-14 (^{14}C)

At the end the question remains: If cosmic radiation is so extremely energetic and so extremely intense, does this mean that the atmosphere, which is supposed to act as a life-protecting shield against this radiation, will itself become radioactive over time? The answer is: "It's unavoidable". During the formation of a particle shower in the atmosphere, up to 10^{11} protons and neutrons are successively released as a result of nuclear reactions. A plethora of residual nuclei are, of course, left behind, and the majority of these are in fact radioactive. Fortunately, these radioactive residual nuclei are predominantly short-lived with lifetimes in the order of seconds to minutes. An appreciable accumulation in the atmosphere or biosphere will therefore not occur. For some longer-lived residues, the formation probabilities are low and biological consequences in the biosphere need not be considered either.

Of a completely different quality, however, is the large number of free neutrons. Neutrons have a $20-50$ times stronger damaging effect on organisms than the ionizing radiation released by radionuclides. The maximum of neutron production by cosmic rays lies at an altitude of about $15-30$ km. Here the time-averaged neutron flux reaches values of about 6×10^6 neutrons/m^2/s. In contrast, the neutron flux produced at the Earth's surface is only about $200-300$ neutrons/m^2/s. In the presence of buildings

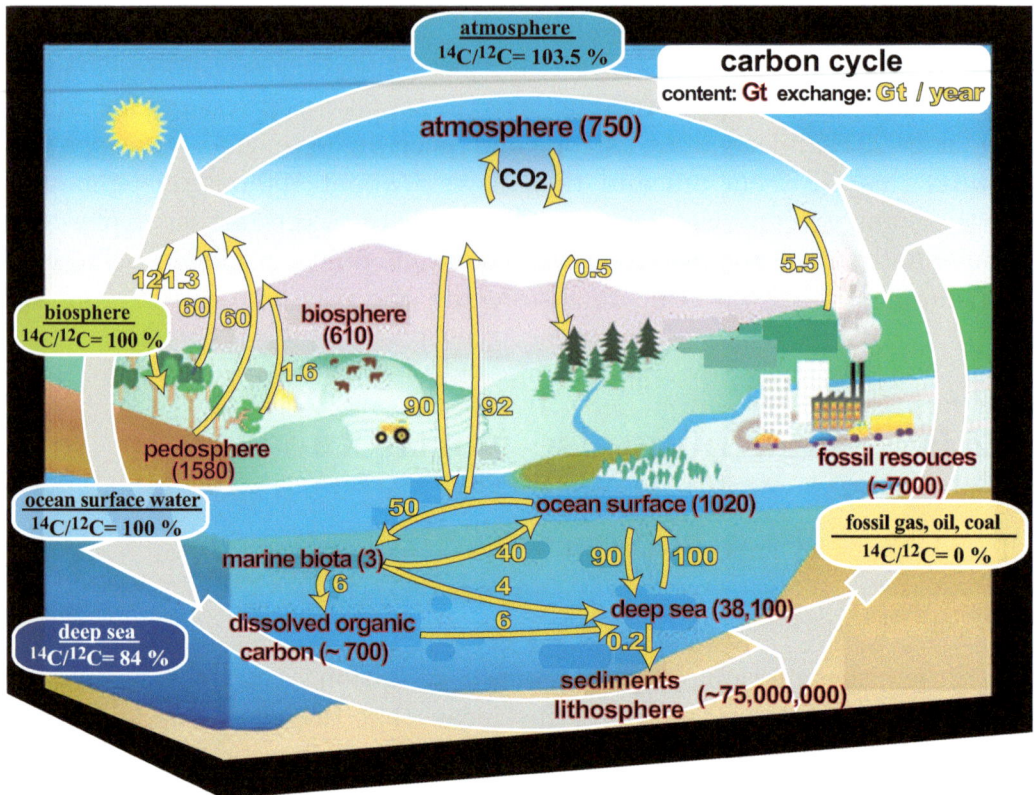

Fig. 12.30: Carbon cycle between the atmosphere and various carbon-containing areas of the Earth. Numbers in red denote the size of each carbon reservoir in Gigatons and in blue the annual amount of exchange in Gigatons. Also plotted are the carbon-14 (^{14}C) concentrations in each area in units of the mean biosphere concentration of 1.3×10^{-12} g/g.

or larger structures, it is somewhat higher because cosmic particle showers impinging here produce some extra neutrons. Further, the flux at the poles is about twice as high as at the equator due to the Earth's magnetic field. In fact, the knowledge of a strongly increasing atmospheric radiation exposure with altitude bears as a consequence that the flying personnel of commercial airlines is required to undergo separate radiation protection monitoring (in Germany carried out by the "Bundesamt für Strahlenschutz", i.e., Federal Office for Radiation Protection).

Interestingly and fortunately, the majority of neutrons produced at high altitudes never reach the Earth's surface. The reason for this is a special and almost singular nuclear physics property of nitrogen. Nitrogen is an extraordinarily effective "neutron absorber", and here it is only the isotope nitrogen-14 (^{14}N), which possesses this property. Conveniently, ^{14}N is also the most abundant isotope in Nature with a natural abundance of 99.63% (the other stable isotope is ^{15}N with an abundance of 0.37%), and with 78%, nitrogen also makes up the largest part of the Earth's atmosphere.

The reaction, which takes place on nitrogen-14 in the atmosphere,

$$^{14}\text{N} + n \longrightarrow \quad ^{14}\text{C} \ + \ p$$
$$\hookrightarrow {}^{14}\text{N} + e^- + \bar{\nu} \quad (T_{1/2} = 5730 \text{ yr}, E = 157 \text{ keV})$$

converts the neutron to the proton — however, co-produced next to the harmless proton is the radioactive carbon-14. Carbon-14 has a long half-life of 5730 ± 30 years and decays back to ^{14}N by means of a β-decay. The decay is critical for biological substances, because if a ^{14}C atom bound in an organic molecule decays, the molecule is suddenly confronted with a nitrogen atom in its chemical structure. The immediate breakup of the molecule is the consequence.

However, the chain for the incorporation of ^{14}C into an organic molecule is a bit more convoluted. The ^{14}C atom formed in the atmosphere immediately combines with oxygen in the atmosphere to form carbon dioxide ^{14}CO$_2$. From there on, it is part of the well known carbon cycle shown in Fig. 12.30, which initially begins with a rapid mutual exchange of CO$_2$ between atmosphere, biosphere, and the total surface water of the world's oceans. The mean exchange times are short compared to the ^{14}C half-life. They are about 10 years, so that ^{14}C formed in the atmosphere is distributed rapidly and uniformly over large quantities of carbon-containing compounds in the biosphere. It is precisely this property that forms the basis for the radiometric ^{14}C dating method.

In the biosphere, the ^{14}C concentration is about 1.3×10^{-12} g/g [24]. In the atmosphere, which is after all the ^{14}C production site, the concentration is about 3% higher, and in the deep ocean water it is about 14% lower because

---

[24] Long-term variations in cosmic-ray flux as well as short- and long-term solar cycles, affect ^{14}C production in the atmosphere and produce deviations from the mean biosphere value, which are of the order of a few percent. Furthermore, the burning of fossil fuels has led to a decrease in ^{14}C concentration over the last 100 years.

of the long exchange times of about 1000 years. In fossil fuels such as coal, natural gas and oil, the exchange times are so long that no accumulation of ^{14}C can occur.

Since ^{14}C is continuously incorporated by living organisms, the average ^{14}C concentration in humans, for example, results in a decay rate of about 4000 Becquerel (decays per second).

In total, cosmic rays impinging on Earth's atmosphere cause a production of about 4 kg of ^{14}C annually. Thus, the equilibrium amount of ^{14}C existent in the entire biosphere, which is given by the balance between production and decay, comes to about 40 tons.

Small addendum: Without this special property of nitrogen-14 as an effective atmospheric neutron filter, the evolution of living organisms on Earth's surface would have been significantly different.

Chapter 13

Escape to the Future

The prospects for planet Earth and its inhabitants don't look good in the distant future. Several events of cataclysmic proportions have already been mapped out. After about 400 – 900 million years the Sun gradually gets hotter, so that the living conditions on Earth become increasingly unfavorable. The Sun remains relentless and at some point in its final phase it will simply swallow the Earth. For the intelligent inhabitants on Earth, this means that another and preferably safer planet must be found as early as possible — somewhere in the Milky Way — and it better be one, which allows a timely re-population. And sure, it ought to be one not already inhabited with intelligence — as this inevitably calls for conflict. But even so, the next catastrophe is around the corner, because the huge Andromeda Galaxy, which can already be seen with the naked eye in the dark sky, is on a direct collision course with the Milky Way at a speed of about 400,000 km/h — meeting point in about 3.5 – 4.3 billion years. If that happens, such a high rate of star collisions and violent star formations will be triggered in the subsequent 3 – 4 billion years during galaxy merging, that it may be advisable to step aside cosmically.

And again the search for a new planet begins, but this time in a different and preferably far away galaxy so that this one won't suffer a similar fate. Nevertheless, in the far distant future, roughly in about 100 billion years, the entire galaxy neighborhood is threatened with an unhealthy proximity to extremely massive Black Holes. Then one would have to move really far away. Since the Universe has now expanded by more than an order of magnitude, a search could be a bit more time-consuming.

But perhaps solving the current problems on Earth is for now a bit more of a priority.

© The Author(s), under exclusive license to Springer-Verlag GmbH, DE, part of Springer Nature 2025
D. Frekers, P. Biermann, *Universe, Neutrinos, Stars and Life*, https://doi.org/10.1007/978-3-662-70729-6_13